No 1 Squadron

Previous page:
Despite spells of appalling weather, the Squadron flew 126 operational sorties during the Falklands War. Here the aircraft are lashed down on the rolling deck of *Hermes* as she sits out a gale to the east of the islands. *Squadron Archives*

Michael Shaw

No 1 SQUADRON

LONDON

IAN ALLAN LTD

Contents

It has been a conscious decision to include in this volume photographs of historic interest which might, in other conditions, have been excluded on grounds of quality. It is hoped that the reader will accept the consequent reduction in the production standards of some plates.

First published 1971;
this edition 1986

ISBN 0 7110 1581 3

Published by Ian Allan Ltd, Shepperton, Surrey; and printed by Ian Allan Printing Ltd at their works at Coombelands in Runnymede, England

Preface

My first attempt at writing the history of No 1 (Fighter) Squadron took place in 1970 under strenuous circumstances. In company with the other pilots of the Squadron, I was struggling to get to grips with the shiny levers of our newly-arrived Harriers, unaided in those days by flight simulators or two-seat trainers. Our new CO, flinty young Wing Commander Ken Hayr, drove us like slaves all day to master the strange ways of our new machines. And in the evenings and nights, I wrote the book.

Small wonder, I suppose, that eventually I bent three Harriers in as many days and was sent on leave with orders to finish the wretched book and get back to work refreshed.

Fifteen years later *Twice Vertical** was out of print and out of date. In the intervening years No 1 had added some stirring deeds to its noble past. On two occasions the unit's timely arrival in Belize has been instrumental in preventing a Guatemalan invasion. Then, in 1982, the Squadron went to war in the South Atlantic, 11,000 miles from home. We, their grounded predecessors, were mightily proud of what those pilots wrought over Port Stanley and Goose Green. I therefore undertook to tart up the old book in their honour, and to place their deeds alongside those of Fullard and Clayson, of Clisby and Hilly Brown.

In so doing I must thank Wg Cdr John Feesey and the present members of the Squadron for their hospitality and help; also MRAF Sir Dermot Boyle, AVM Ken Hayr, Wg Cdr Eric Smith and the many members of the No 1 Squadron Association who lent me their photos and their memories to make this book possible. Also Alfred Price, Roger Darlington and the staffs of the Air Historical Branch and the RAF Museum. And finally Pippa, my wife, typist and critical proof-reader, who, in the middle of it all, gave birth to our son. This book belongs to you all.

Michael Shaw

**Twice Vertical — The History of No 1 Squadron* was published by Macdonald & Co in 1971.

CHAPTER ONE

Vertical Prelude

The Royal Flying Corps was formed on 13 May 1912. A vital link was forged that day in the chain of events that led to the establishment of the Royal Air Force. Yet for the senior unit of the new Corps, this was no beginning: No 1 Squadron, Royal Flying Corps, already had a distinguished pedigree that stretched back unbroken to the very beginning of military aviation in Great Britain.

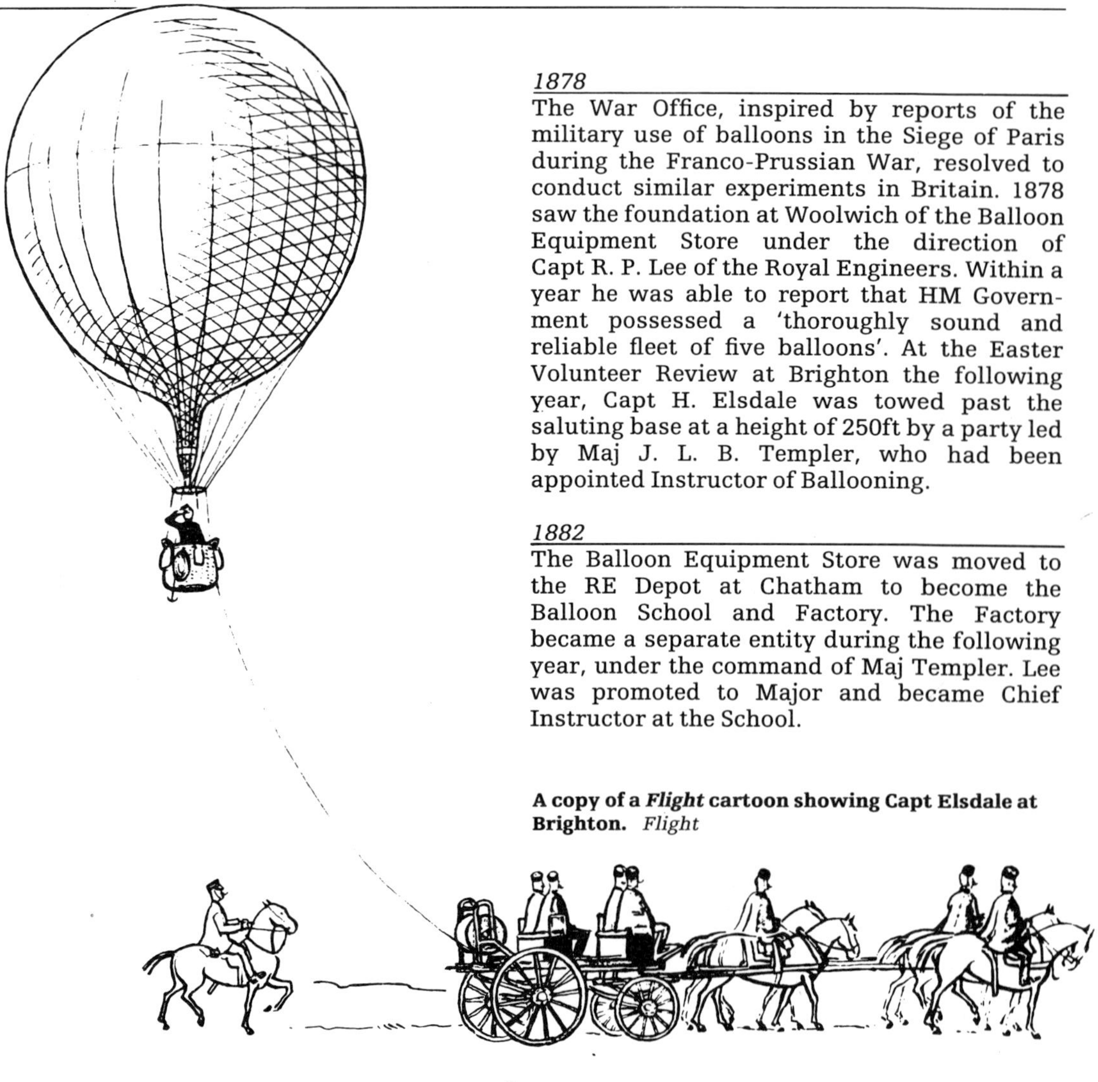

1878

The War Office, inspired by reports of the military use of balloons in the Siege of Paris during the Franco-Prussian War, resolved to conduct similar experiments in Britain. 1878 saw the foundation at Woolwich of the Balloon Equipment Store under the direction of Capt R. P. Lee of the Royal Engineers. Within a year he was able to report that HM Government possessed a 'thoroughly sound and reliable fleet of five balloons'. At the Easter Volunteer Review at Brighton the following year, Capt H. Elsdale was towed past the saluting base at a height of 250ft by a party led by Maj J. L. B. Templer, who had been appointed Instructor of Ballooning.

1882

The Balloon Equipment Store was moved to the RE Depot at Chatham to become the Balloon School and Factory. The Factory became a separate entity during the following year, under the command of Maj Templer. Lee was promoted to Major and became Chief Instructor at the School.

A copy of a *Flight* cartoon showing Capt Elsdale at Brighton. *Flight*

The balloon camp at Lydd in 1908. This very early example of aerial photography was taken by Maj H. Elsdale.

1884

The first active service by British military balloons took place in Bechuanaland. Maj Elsdale and Capt F. C. Trollope led a section of three balloons as a part of Sir Charles Warren's expedition. The balloons had difficulty getting airborne at the high elevation of the African veldt, and no fighting was actually observed from the air. The local population, however, was reported to be greatly amazed by this display of modern military science.

1890

After a period of financial uncertainty Sir Evelyn Wood reported favourably to the Treasury on the value of military ballooning. This resulted in the formation of the Balloon Section of the Royal Engineers. The Balloon Depot remained at Chatham, while the Balloon School and Factory moved to a new site on Laffan's Plain near Aldershot, called Farnborough. The Balloon Factory later became the Royal Aircraft Factory which developed into the Royal Aircraft Establishment of today. The Balloon depot followed across to Farnborough in 1892, and the School of Ballooning was later established as a separate entity.

1899

The Balloon Section sent four detachments to South Africa to serve in the Boer War. During 1900 two of these detachments saw active service at the siege of Ladysmith, and one other at the battle of Magersfontein. Col Arthur Lynch reported to the War Office that the Boers were 'much disquieted by the presence of the balloons . . . when employed in spotting enemy strategems and ambuscades'.

A balloon section on the road in 1895. Lt C. F. Arden-Close of the Balloon Section RE recorded: 'The wagons were towed by a steam sapper at 4-5mph. The CO rode ahead in patrol jacket, riding breeches, jackboots and sword, followed by a sapper carrying a red flag in accordance with the road regulations.'

The balloons were also used for directing gunfire, and one detachment experimented with a pair of very early Marconi wireless sets for this purpose. Unfortunately, the induction coil of the set gave off a fat spark when transmitting. This was deemed to pose a grave danger to the gas-filled balloon, so the observer had to resort to the established techniques of semaphore and heliograph.

One of the Section's balloons on exercise on Salisbury Plain. *Squadron Archives*

1902

Work began on the design and construction of Britain's first steerable airship or 'dirigible'. This was the *Nulli Secundus*, powered by a 45hp Antoinette engine. The project brought together many of the country's emerging designers and engineers. The young Geoffrey de Havilland began his distinguished career by designing the special radiator.

Right:
The balloon *Swallow* moored during manoeuvres, and protected from the wind by canvas screens. Balloons like this had a capacity of about 10,000cu ft. The fabric panels of the envelope were held together with 'gold-beaters skin', produced from the intestines of oxen.
Squadron Archives

Bottom:
The first British airship, *Nulli Secundus*, flown by C. F. Cody, taking off from Farnborough for her maiden flight over London on 5 October 1907.
Squadron Archives

1907

Owing to lack of funds and experience, it was not until 1907 that the airship was handed over to the School of Ballooning for its first flight. This took place in a blaze of publicity on 5 October, with the legendary C. F. Cody in command. Just before lunch, the largest machine yet to have graced the English skies lifted off and clattered away towards London, where she cruised over Buckingham Palace and St Paul's before setting course for home, having delighted and intrigued the capital. Unfortunately, a westerly wind had now risen, in whose teeth she could not progress. So instead of a leisurely trip home to Farnborough, she battled her way to Crystal Palace by dusk, where Cody was able to land on the cycle-racing track. The airship was then dismantled and returned to Farnborough on lorries.

After a further flight she was enlarged from 56,000 to 85,000cu ft, but only made two further flights before being scrapped. Much more was now known about the basics of airship design, and the lessons of *Nulli Secundus* were put to good use in the construction of the family of British dirigibles that now followed.

1908

Serious experiments were under way with wireless communications from balloons, under the direction of the new superintendent of the School, Col J. E. Capper. These were conducted with great zeal by Capt Llewellyn-Evans of No 1 Wireless Company RE, despite the fact that he was usually airsick when aloft. During May, Col Capper and Lt C. J. Aston, cruising over Farnborough at 8,000ft in the balloon *Pegasus*, received signals both from

The airships *Beta* (airborne) and *Gamma*, taking part in the Army manoeuvres near Cambridge in 1912. *Squadron Archives*

Aldershot and from a warship lying at Portsmouth. In the following month, Lt Aston and Capt King managed to transmit a recognisable message during a trip from Aldershot to Godalming, where they were forced to land by a thunderstorm.

1910

Two new airships — *Beta* and *Gamma* — now emerged from the factory. *Beta* became the most successful of the early British dirigibles, making a total of 331 flights before being withdrawn in 1916. Her first flight was made after dark, because the air was much calmer and it was considered at the time that there was 'more lift' in the air at night. She was originally powered by two 8hp Berliet engines which were quite insufficient, giving *Beta* a forward speed of 2kt! She was modified several times and had the distinction of carrying the Prince of Wales on his first flight on 20 July 1912. The gondola in which he rode can be seen today at the Science Museum in South Kensington.

Gamma was less popular with her pilots than *Beta* due to some disturbing aerodynamic habits that made her difficult to steer, despite many modifications. Her nickname of 'The Yellow Peril' was thus only partly due to her colour.

1911: The Air Battalion

By now the size of the ballooning establishment was fast outgrowing its status as a Section of the Royal Engineers, and an increasing amount of work was also being done with aeroplanes. To rationalise the situation, the War Office ordered the creation on 28 February 1911 of the Air Battalion, Royal Engineers. This is generally accepted to have been the first properly constituted military flying unit in the world.

The Balloon School, having control of all airships, balloons and man-carrying kites, became No 1 Company of the Air Battalion and remained at Farnborough, while No 2 Company moved to Larkhill and assumed responsibility for the aeroplanes.

The task of the Air Battalion was to 'create a cadre of expert airmen to facilitate the formation of units ready to take to the field with troops, and be capable of expansion by reserve forces'. No 1 Company, having assumed the mantle of the Balloon School, embarked on a programme of expansion by attracting officers and men seconded from other Army units. The qualities sought when selecting aspiring airmen varied from 'previous aeronautical experience' to being a 'good map reader and field sketcher'. It was added hopefully that applicants needed to be good sailors.

Maj E. M. Maitland — the Commanding Officer of No 1 Company of the Air Battalion, RE, and later the first CO of No 1 Squadron, Royal Flying Corps. *Squadron Archives*

A mechanic, suspended from a balloon, repairs the upper valve of airship *Gamma*. *Squadron Archives*

Major E. M. Maitland commanded No 1 Company from the outset. He had become famous for a world record balloon flight in 1908, in company with Maj C. C. Turner and Professor A. E. Caudron, flying in the huge balloon *Mammoth* from Crystal Palace to Central Russia, covering 1,117 miles in 36½ hours.

The Chief Instructor was Capt C. M. Waterlow RE, a quiet man with a wide knowledge of aerostatics who commanded *Gamma* with great skill. Little *Beta* was commanded by Lt J. H. Fletcher RE, who was completely devoted to flying and to aeronautical science. His second officer in *Beta* was Lt the Hon J. D. Boyle, seconded from the Rifle Brigade.

Under the watchful eyes of these more experienced officers, several carefully selected 'balloonatics' made swift progress until it was decided at the War Office that the driving of flying machines was an undignified pastime best left to a competent NCO, and that officers should be trained as observers and left free to command the machine and its reconnaissance from the back seat. Maitland fought this policy from the outset, rightly insisting that the commander of any aircraft should be its pilot. A group of aged staff officers appeared for training as observers, but fortunately the training facilities were so limited that the War Office was obliged to drop the idea.

1912: The Royal Flying Corps

The Government was now becoming increasingly worried about the pace of development of air power overseas. France was leading the world in aeroplane design, while Germany, with whom relations were becoming strained, was building a fleet of large airships designed by Graf von Zeppelin.

The committee of Imperial Defence had been watching these developments, and in April 1912 it accepted a scheme for establishing an autonomous military flying service of Corps status. The approval of King George V led to the formation on 13 May 1912 of the Royal Flying Corps — the world's first separate air arm.

The RFC, although responsible entirely to the War Office, was initially established with a Military Wing and a Naval Wing, although the latter was subsequently taken over by the Admiralty to form the Royal Naval Air Service. The composition of the Military Wing included the Wing Headquarters, one airship and kite squadron, seven squadrons of aeroplanes and the RFC workshops at Farnborough.

The two companies of the old Air Battalion naturally became the basis on which the Military Wing of the RFC was built. Now independent of its Royal Engineer ancestry, the new Corps adopted the cavalry title of Squadron for its fighting units. No 1 Company of the Air Battalion thus became No 1 Squadron of the Royal Flying Corps on 13 May 1912.

CHAPTER TWO

First Blood

No 1 Squadron first took to the field to participate in the army manoeuvres around Cambridge in July 1912. The airships *Gamma* and *Delta* were equipped with wireless, but *Delta* broke down and was replaced with little *Beta*. One airship was attached to each side and valuable experience was gained in providing reconnaissance information for the Army.

Later in the year a French airship designed by Clement-Bayard, and later named *Zeta*, was added to the fleet, followed by *Eta*. More officers were posted in for training, and the nominal roll began to read like a page from Debrett. Maj the Hon C. M. P. Brabazon of the Irish Guards became the second officer of *Delta*, in which he had to handle the swivelling propellers that enabled her to lift off at a weight of 500lb heavier than air. Lord Edward Grosvenor was seconded to No 1 from the Life Guards. He was the son of the Duke of Westminster, and used to race a large Mercedes at Brooklands with Lt T. G. Hetherington when off duty.

The manoeuvres of 1913 were conducted for the first time on a full war footing, and training continued with a new urgency as the international situation deteriorated. Various types of experimental parachute were investigated, and the courageous Maj Maitland leaped from the gondola of *Delta* to make the first parachute jump from an airship.

On 1 January 1914, control of all airships was passed to the new Royal Naval Air Service, and over the next few months they were

Airship *Delta*. The swivelling propellers can clearly be seen: they enabled the airship to lift off at a weight of 500lb heavier than air. *Squadron Archives*

Above:
The officers of No 1 Squadron, during the Army manoeuvres of 1913. Left to right: Lt J. D. Mackworth, Maj the Hon C. M. P. Brabazon, Mr Wilson (Royal Aircraft Factory), Lt the Hon J. D. Boyle, Lt T. G. Hetherington, Lt Lord Edward Grosvenor, Capt C. M. Waterlow, Maj E. M. Maitland, Lt J. N. Fletcher, Capt R. Pigot, Capt H. P. T. Lefroy, Capt Vidal (RAMC), Capt O'Reilly. *Squadron Archives*

Maj Maitland making the first parachute jump from an airship, on 6 October 1913. *Squadron Archives*

dispersed to the various coastal air stations. With them went Maj Maitland, who later became head of the Airship Branch during the war, when he would drop in on various units by parachute, followed by his batman and suitcase! He was eventually lost with the R38 which crashed in the Humber in 1921.

All that remained of the Squadron was Major Brabazon's flight of man-lifting kites and a small administrative cadre. Then even the kites were removed to form No 6 Squadron.

For a while, No 1 lay in limbo, until it was decided to re-form the unit as an aeroplane squadron in the reconnaissance role. On 1 May Maj C. A. H. Longcroft was posted in from No 2 Squadron to command. No aeroplanes were forthcoming at first, and it was with only a small nucleus of officers and men that No 1 Squadron went to war on that fateful 4th of August.

At the outbreak of war, Nos 2, 3, 4 and 5 Squadrons of the RFC were ordered swiftly to France, and it was to the dismay of the cadre of No 1 Squadron that Maj Longcroft was hurriedly posted back to his old unit. Several other pilots followed him to the front and the remainder were further frustrated to hear that their old CO was already in action, spotting the advance of German forces on 17 August.

Acting command of the Squadron passed to Capt F. V. Holt and on 15 August it moved to Brooklands. Later in the month the first aeroplanes began to arrive. By early September a total of a dozen machines had been received, including two Bleriot monoplanes, three Maurice Farmans, a BE8 and a Morane monoplane. Maj Longroft then returned from France and training was begun to bring the Squadron up to operational status.

On 13 November, during this period of intensive training, No 1 moved to Netheravon to join No 7 Squadron in forming the 4th Wing RFC, under Lt-Col J. F. A. Higgins, who had also seen service in France in command of No 5 Squadron. Higgins was universally known in the RFC as 'Colonel Bum and Eyeglass' because of his jutting backside and perpetual monocle. Training continued with the addition of officers posted in as observers. During December, a number of aircrew were detached to become the nucleus of No 12 Squadron which was in the process of formation.

Maj Longcroft returned to France on 28 January 1915, to take over No 4 Squadron; he was succeeded in command by Maj W. G. H. Salmond. Geoffrey Salmond arrived from France where he had been on the staff of Brig Sir David Henderson at RFC Headquarters, and with him he brought news from the

Below:
A Bleriot monoplane at Netheravon in the summer of 1914. *Squadron Archives*

Bottom:
No 1 Squadron used three Maurice Farman Longhorns like this one for training before departing for France in March 1915.

Right:
A BE8 of the type used on the squadron's first operation on 12 March 1915. *Squadron Archives*

Below:
Bailleul aerodrome in 1915, showing the canvas hangars and the boarded-over pond. The lunatic asylum is in the background.
Squadron Archives

front line of how the early mobile fighting had begun to settle down into trench warfare. The RFC squadrons already in France were meeting German aircraft high above the opposing armies and were attempting to deny their reconnaissance by driving them off with rifles. His words fell on eager ears. The following month notice arrived of No 1 Squadron's forthcoming move to France.

During February, four BE8s and eight Avro 504s were received and the ponderous old Farman 'Rumpitties' were relegated to the Flying Training Schools. Final preparations were made, goodbyes were said and the ground support party and transport left England on 3 March.

The 12 aircraft and crews left Netheravon for Folkestone on 7 March. They crossed the Channel without incident, and were reunited with the groundcrew at St Omer the following day. On arrival, they discovered that they had been attached to the 3rd Wing, RFC, commanded by Lt-Col H. R. M. Brooke-Popham. They were met by Longcroft's No 4 Squadron, with whom they were to form the Wing.

There was little time to settle in to their new surroundings, for their arrival coincided with the opening moves of the Battle of Neuve Chapelle, and the Squadron flew its first operational reconnaissance within 48 hours of arriving in France. On 12 March the first offensive sortie was flown. Four aircraft, flown by Capt Edgar Ludlow-Hewitt, Lts V. A. Barrington-Kennett, Eustace Grenfell and O. M. Moullin, took off in an attempt to destroy a rail bridge to the northeast of Douai and a railway junction at Don. 20lb bombs were dropped over the side by the pilots, because observers were not being carried, but no hits were observed. Worse still, Lt Moullin failed to

An Avro 504 in front of 'B' Flight's hangar at Bailleul. *Squadron Archives*

return. He was later reported to be a prisoner of war.

When the Battle of Neuve Chapelle died down, the Squadron settled into learn the area and to train under conditions of active service. Wireless and photographic co-operation with the army was extensively practised, and on 29 March the Squadron moved to Bailleul and transferred to the 2nd Wing which was in support of II Corps around Ypres.

Unlike many RFC squadrons, which seemed to be perpetually on the move, No 1 was to remain at Bailleul for exactly three years. It was a pleasant aerodrome, located just outside the town whose name it bore. Only 11 miles west of the front line at Ypres, Bailleul was close enough to the war to guarantee continuous action for its squadrons. At first, No 1 had the aerodrome virtually to itself and was accommodated — some unkind souls said appropriately — in a large lunatic asylum on the edge of the field. The three flights of the Squadron were hangared in front of the asylum, and in the centre of the airfield was a large pond: Maj Salmond ordered this to be boarded over following an unfortunate and rather damp taxying incident. The arrival of other units caused a lot of overcrowding in the asylum and the groundcrew had to be moved out to a large barn on the outskirts of the town, from which they marched every morning for breakfast and the day's operations. Shortly after No 1's arrival, a Zeppelin dropped some bombs on the aerodrome, without causing any casualties.

At the beginning of April the Squadron found itself faced with increasing reconnaissance and artillery observation duties. Just south of Ypres lay the enemy-held feature known as Hill 60, on which a major assault was planned for 17 April. Tunnels had been dug from the Allied trenches and mines were laid under the hill. Reinforcements were brought up, and No 1 was given the additional task of keeping the sky above the area free of German reconnaissance machines — the Squadron's first counter-air task.

The whole front from Kemmel to Ypres was covered that day by continuous standing patrols from 04.30 until 19.15hrs. Additional reconnaissance patrols were flown to spot and report enemy artillery positions by the flashes of their guns. Capt Ludlow-Hewitt also flew a special 'flash recce' to assess the actual detonation of the mines. The following even-

748

Left:
Martinsyde S1 No 748 at Bailleul. This aircraft, flown by Lt Micky Bell-Irving, was involved in one of the Squadron's first engagements with a German aircraft, on 11 May 1915. *Squadron Archives*

Below left:
A Morane Parasol being 'Archied' over the trenches. *Squadron Archives*

ing, the flashes of 33 enemy guns were spotted by eight aircraft led by Maj Salmond. Throughout the action, the concentration of our troops went completely unobserved by the enemy, and the attack thus came as a complete surprise. Although the fighting for Hill 60 died down by 21 April, the enemy launched his first gas attack of the war on the following day. This did not directly affect Bailleul, but it was later considered necessary from time to time for everyone to wear 'flannel bag' gas clothing, which was to cause a lot of discomfort on the hot summer days ahead.

On the 29th, the Squadron was once more attached to the 3rd Wing, but continued to operate from Bailleul, which it now shared with No 4 Squadron. At this time No 1 was flying a rather mixed bag of aircraft, and in early May possessed three Avro 504s, four Morane Parasols, one Martinsyde and a remaining BE8. The new Morane was of French design and could be longitudinally unstable at times, but its high monoplane wing, strung from a kingpost protruding up from the forward fuselage, gave it an outstanding downward field of view which made the aircraft very suitable for reconnaissance.

The art of air combat had not yet been developed into the hectic dogfighting of later years, and successes at this early stage of the war were comparatively rare. Reconnaissance machines patrolled singly as a rule, busy with their tasks of army co-operation, but would engage enemy aircraft with their Lewis gun if any appeared. The gun was usually operated by the observer from a swivel mounting. This called for a good understanding of the art of deflection shooting, especially when standing unstrapped in a cold slipstream.

The BE8 was being phased out of service. Much loved by its crews, it had earned the nickname of 'Bloater', and it was in the final days of its service with No 1 that the last BE8 took part in the first combat recorded by the Squadron. On 3 May, Lt V. A. Barrington-Kennett and his observer, Lt F. L. Richardson, spotted an enemy Aviatik over Lizerne. They pursued it to the Houthhulst Forest, but were unable to catch up, so climbed to 7,000ft and were able to pounce on it when it turned. Richardson fired some 35 rounds at it from above with his Lewis gun at close range. The German disappeared in a vertical dive towards the forest. No positive claim was allowed, as the aircraft was not seen to crash.

Increasing enemy activity over the front produced a number of inconclusive engagements during May, but it was not until 5 June that Capt C. E. Stodart, regarded by the ground crew as a 'bit of a mad blighter', could claim the destruction of a German aircraft short down by his observer, 2nd Lt M. S. Stewart, with their Avro 504.

In the period between their arrival in France and 20 June 1915, the Squadron also flew 141 bombing sorties against railway junctions and bridges behind the enemy lines. Bombing was so inexact at this time, however, that only three of these produced any tangible results. Bombs were dropped hapazardly over the side by the observer without any form of sighting apparatus, and it was not until the later introduction of bomb racks and dive bombing that the considerable inaccuracies were reduced. An assortment of improvised bombs, ranging from mess rubbish to inscribed chamber pots, was also dropped as the mood took the crews, as well as clusters of little steel darts called 'Flechettes', which were later banned because their indiscriminate dropping was likely to kill innocent civilians.

The early summer of 1915 saw the introduction of the first machines specifically designed for aerial combat. The Germans were first in this field with the Fokker EI monoplane and the early models of Albatros, which were fitted with Parabellum machine guns that fired forward through the propeller disc by means of an interrupter device. This was a significant development as it enabled the pilot to aim his entire aircraft at his opponent. German pilots such as Oswald Boelcke and Max Immelmann soon developed the art of deflection shooting in the air, and began to take a heavy toll of the ponderous Allied two-seaters.

The enemy anti-aircraft fire — or 'Archie' — was also becoming more accurate, and No 1 Squadron had suffered its second casualty of the war on 23 May, when Lt C. Court-Treatt was wounded. On the 24th, 2nd Lt H. F. Boles was wounded in an engagement with one of the new Fokkers, and died of his wounds the following day. Soon after this, Lt M. McB. Bell-Irving was awarded the DSO for his tenacious action in a fight with three Fokker monoplanes. Micky Bell-Irving came from Vancouver and was one of three brothers serving in the RFC. On the day in question, he was flying solo in a Morane that was not fitted with a Lewis gun, and had taken a rifle with him for protection. When the Fokkers attacked, he dived for the safety of friendly territory, unstrapped and opened fire on the enemy, eventually driving them off. Bell-Irving was later chosen to provide the flying display when Lord Kitchener visited the Squadron on 18 August 1915.

Defensive measures had to be taken to offset the advantages of the enemy scouts, and some of these resulted in the adoption of tactical formations used to this day. When flying alone, however, greater ingenuity had to be employed to survive. 2nd Lt R. Balcombe-Brown and Lt M. C. K. Hughes were airborne on reconnaissance between Ypres and Courtrai one day when they were bounced by four Albatros Scouts. Their Morane Parasol was armed with a fixed forward-firing Lewis gun on the upper surface of the wing, which could not be brought to bear on the pursuing Germans. Hughes, however, whipped out his spotting telescope and pointed it at the nearest Albatros, which had closed to some 250yds, whilst his pilot made appropriate firing noises with the Lewis. This caused the German to sheer off and follow the Parasol at a more discreet distance until it crossed the lines, beyond which the enemy machines did not pursue the matter further.

Although the enemy scouts rarely crossed into Allied airspace, they were causing heavy losses to most of the RFC two-seater squadrons. Such were the Allied casualties in the summer of 1915 that the period was known to many as that of the 'Fokker menace'. When Brigadier H. M. Trenchard, later to be the founder of the Royal Air Force, took command of the RFC in the summer of 1915, the situation was very bleak. The Germans had clearly established air superiority over their own territory. Public opinion brought pressure to bear on Westminster to reduce air losses, and the new commander was instructed to despatch fewer flights behind enemy lines. This was steadfastly resisted by Trenchard, who rightly insisted on his policy of carrying the war to the enemy. Failure to do so would have deprived the British Army of its most valuable source of reconnaissance.

So it was at a time of some adversity that Maj Geoffrey Salmond handed command of No 1 Squadron to Maj P. B. Joubert de la Ferté. The new CO, later to become ACM Sir Philip Joubert, KCB, CMG, DSO, before his retirement in 1947, took over the reins on 15 August, while his unit was supporting the 2nd Army at the attack on Hooge.

The Battle of Loos began on 25 September, and the squadron was fully engaged in support of II Corps around Lille, patrolling the lines between Habourdin and Comines. During the battle, crews took part in an interesting trial in the use of Brock Flares, which were made by the well known firework firm of the same name. These were lit by troops in forward positions to identify their location and prevent friendly reconnaissance aircraft calling down artillery strikes on them. The flares proved easily visible from 6,000ft, and the trial led to their general introduction on the Western Front.

Summer faded into autumn, and the fields of Flanders deteriorated into a morass of mud and shattered mankind, into a quagmire of wartorn woods and villages, into a soulless winter. With the worsening weather, illness began to side with shellfire to take its toll of men. One casualty was the Squadron Commander, who was taken to hospital with severe influenza. Maj Joubert thus incurred the wrath of the stoical Trenchard, who considered that succumbing to illness was a grave display of human weakness. This fall from grace was not to last, for Joubert was later to rise to the highest ranks under Trenchard's command. His place was taken on 24 November by Maj G. F. Pretyman, who had been with No 3 Squadron at the retreat from Mons in the very early days of the war. An experienced and brave pilot, he was an old friend of Maj Joubert and was able to step into command without difficulty, having been advised by his predecessor that he would find No 1 Squadron's crews 'a splendid lot'. Shortly before Joubert's departure, Lt Eustace Grenfell shot down an Aviatik and two other enemy aircraft on the same day.

And Christmas was coming. In spite of food shortages at home, several national newspapers had run subscription funds to provide the men at the front with a little parcel to remind them that they had not been forgotten. Christmas puddings, sweets, chocolate and cigarettes were sent, and the officers and men of No 1 Squadron did their best in damp and dismal surroundings to forget the war for a day. A concert party was organised, with great rivalry between the Flights to provide the best act. The beer flowed freely, and jealously hoarded supplies of the rum ration were broken out. Even Sgt Maj Darke, in charge of discipline on the unit, was occasionally seen to smile.

The Morane Biplane was mainly used for the dangerous 'long reconnaissance' sorties of 1916.

The year of 1916 opened with a series of 'long reconnaissance' flights for the 2nd Army. To provide the endurance required for these long, cold and hazardous sorties, the Squadron received two of the new Morane biplanes. Laden with fuel, these aircraft would cross the lines alone to plot troop and artillery movements deep inside enemy territory. Their vulnerability to the roving Fokkers soon became evident and on 11 January Lts J. C. McEwan and F. Adams were shot down, but lived to become prisoners of war. The following day the other Morane was reported missing on a similar sortie, and Lts R. Barton and E. S. Wilkinson were later presumed killed — an inauspicious start to the year. Losses of machines from other units engaged in long reconnaissance were even heavier. On 14 January RFC Headquarters ordered that future sorties were to be escorted and that 'owing to the superiority of the Fokker monoplane, escorted and escort aircraft are to fly in formation'. In spite of these precautions, the Squadron lost a Parasol and its crew, 2nd Lts Watts and Hayward, on the 17th. Replacement aircraft and crews were not long in arriving, and by the end of January, four new Morane biplanes and eight of the old Parasols were on charge.

Until now the Squadron had been part of the 2nd Wing RFC under Lt-Col J. M. Salmond, whose main task was the tactical support of II Corps. He also had additional direct responsibilities to the 2nd Army, which often resulted in a degree of confusion at Wing HQ. The 2nd Army requests for patrol and long reconnaissance, essentially strategic in content, often conflicted with the main day-to-day tactical job of supporting II Corps HQ in the main role of the 'art obs' and close reconnaissance. The resultant over-tasking of wings throughout the Flying Corps had produced a need for additional squadrons, which in turn necessitated an intermediate formation between RFC Headquarters and the Wings. The outcome of all this was the introduction of RFC Brigades early in 1916. Each Brigade consisted of two wings, one directly to support the army to which the Brigade was allocated, and the other for tactical work with the corps in that particular area.

The 2nd Wing, comprising 1, 5, 6 and 7 Squadrons, now became the 2nd (Corps) Wing, and together with the new 11th (Army) Wing came to form No 2 Brigade RFC, which was, logically, allocated for duty with the 2nd Army and II Corps. As No 1 Squadron was now a part of an exclusively Corps-orientated Wing, its future tasks were limited to tactical artillery observation, close reconnaissance and photography. Much to the relief of the Squadron this spelt the end of 'long recce' and produced an immediate reduction in the casualty rate.

In March the Squadron became one of the first two RFC units to receive the new single-seat Nieuport Scout. Known to all as the 'Baby', this delightful and manoeuvrable little aircraft soon became very popular with its pilots. The type had been issued to the Squadron to fly escort to the Moranes, but the unit's observers were rightly apprehensive that the advent of the Scouts would eventually put them out of a job with No 1. This did not happen for some time, as only four Scouts were received initially owing to production difficulties at the French factory. In addition, the Squadron's role was still one of army co-operation, and the need for specialist fighter squadrons had not yet become apparent. The Nieuport 'Baby' was powered by a 110hp Le Rhône rotary engine which gave it a spritely performance after the sluggish Moranes, and the pilots felt that at last they might have the edge over the Fokkers.

The first single-seat scout flown by No 1 Squadron was the Nieuport 'Baby' in 1916. *Squadron Archives*

But the Fokker was still the dominant aircraft in the spring of 1916, and in spite of new defensive tactics and the efforts of the escorts, many RFC reconnaissance aircraft were lost over the Ypres front. On 10 April, Lt-Col D. S. Lewis, now commanding the 2nd Wing, was shot down and killed when flying a No 1 Squadron Morane with Capt A. W. Gale. Such losses of experienced leaders could not be afforded, and Trenchard reluctantly ordered that Wing and Squadron Commanders were not to fly within five miles of the front line. For men like Maj Pretyman, accustomed to flying into action at the head of his men, this was an order that grated hard. On many RFC Squadrons it was also one that was often tacitly to be disregarded over the hard months that lay ahead. No fit and worthwhile CO could easily shrug off the burden of sending his young men out to do mortal battle with the fast and confident Fokkers, while their Major sat behind his desk, stifled in an administrative post. Many of those young men did not come home: 2nd Lt R. P. Turner had failed to return on 3 March, Capt R. A. Saunders was killed on the 14th, and 2nd Lt C. E. Foggin was severely wounded on 27 April.

Micky Bell-Irving was in the wars again. A number of experimental weapons were being tried out during April, including a fearsome device called the Rankin Dart. He was supervising the fitting of one of these to a Nieuport when it exploded without warning. The armourers miraculously survived unscathed, but Bell-Irving was injured by a flying splinter and taken to hospital. He was soon released, and although he came straight back to Bailleul, he had to spend a few weeks on the ground before being permitted to fly operationally. He was later severely wounded on a photographic sortie in a Parasol but brought it home to a safe landing, and was awarded the MC for his gallant conduct in saving his aircraft and observer.

The weather now improved as spring turned into early summer, and the bitter winds no longer froze the fingers of the groundcrew with the same ferocity: throughout the winter they had been obliged to drain the engine oil from the aircraft between flights to prevent thickening. Their many tasks were not easy, for problems were still rife with lack of supplies, spares and replacement aircraft. Criminal penny-pinching was commonplace in contracts placed with some parts of both British and French aircraft industries, and many RFC men had gone to their deaths directly as a result of weak and shoddily-produced equipment. In the face of this, No 1 Squadron's groundcrew became masters of improvisation, and few sorties were lost as a result of unserviceability. The Squadron's 'ack-emmas' lived under very spartan conditions. As the weather improved, many of them moved out of their barn and in the back of the hangars built bunks for themselves from the remains of wrecked aircraft. Their food had been quite appalling until Maj Pretyman authorised an extra messing fund to supplement their humble rations of bully beef and biscuits. Sauces, pickles and local produce were bought to improve their meals, and soon a canteen was opened at Bailleul as well as a primitive forerunner of the NAAFI, where extra food, beer and cigarettes could be bought with their very meagre pay. In spite of the conditions, the morale and teamwork were excellent, for officers and men alike felt that they belonged to the big No 1 family. By comparison with the poor blighters who trudged endlessly past the aerodrome towards the grumbling gunfire of the trenches, they were fortunate indeed.

Throughout those early summer days, plans were being laid for the great allied summer offensive on the Somme — the operational plan that would casually trade half a million British lives for a few miles of shattered French landscape. It was decided that No 1 Squadon's tiny but important role would be to help to eliminate the enemy reconnaissance kite-balloons before the vast troop movements betrayed our detailed plans to the enemy.

Experience had shown that the German *Drache* kite-balloon was relatively invulnerable to conventional machine-gun fire, which merely punctured the canopy and failed to ignite the 27,000cu ft of gas. A patrol from No 1 Squadron had recently tried to shoot down a passing Zeppelin without success, and a punctured balloon could be winched to safety with impunity, so it was decided to use the new Le Prieur *Torpeille* to destroy them. This weapon was the earliest example of airborne rocketry. Invented by Lt Y. P. G. Le Prieur of the French Navy, it consisted of a 4½ft rocket, inclusive of warhead, clipped in salvoes of three or four to the interplane struts of scout aircraft. The rockets were connected to a battery in the cockpit and fired with a tapping key that completed an electrical circuit. The warhead, an adapted incendiary bomb was impact-fused to puncture the balloon and ignite the escaping gas.

On 25 June No 1 Squadron's four Nieuport Scouts were armed up and made ready for a dawn take-off. Simultaneous attacks were to be made on *Drachen* on all fronts of the British armies. The first sorties proved that this was to be no picnic. The rockets had an accurate range of only 300ft, beyond which gravity drop curved the trajectory to the extent that sighting was impossible. The attack involved flying the scout at the balloon until it filled the pilot's field of vision, firing the salvo, and pulling smartly over the top of the ensuing conflagration. To add to the dangers of colliding with the victim, the balloon sites were surrounded by weapon pits filled with light machine-guns, and often with heavier Archie.

The Nieuports were flown all day on balloon sorties, at great expense of rockets and adrenalin, and by evening Lts J. D. Latta and B. J. Moore had each claimed a kill on the 4th Army front. The following day they were at it again. Latta and Moore repeated their previous day's success and Capt R. Balcombe-Brown got one at Quesnoy.

On these two days the RFC destroyed nine kite balloons over the whole allied front, of which five were claimed by a single squadron for no loss. No 1, using just four aircraft to achieve this success, could feel justly proud, especially as Trenchard had let it be known that he considered the destruction of a *Drache* to be worth the loss of three aeroplanes.

Still equipped with a mixture of Parasols, Morane Biplanes and 'Baby' Nieuports, the Squadron prepared for the main Somme offensive. The rumble of thousands of heavy guns far to the south heralded the dawn of 1 July. From the opening of the battle, the RFC kept up a fierce assault behind the enemy lines in an attempt to prevent reinforcement by harassing lines of communication. This was heavily resisted by the German scout units, and fighting in the air increased to the extent that it became impossible to tally individual combats. For the first time, great aerial battles began to supersede individual combat, as escort clashed with defender high above the plodding bomber and reconnaissance two-seaters. Continued requests for observation and reconnaissance meant that No 1 flew escorted sorties from dawn to dusk.

To enable the Squadron to meet its extra load the establishment had been raised just before the battle from 12 aircraft to 18, with the additional machines forming a third flight. Unfortunately, the newcomers were all old Moranes, as new equipment was still not forthcoming. To make matters worse, the Squadron had to make do for some time without additional manpower, and the hard-working groundcrew now found themselves pushed even closer to the limit.

The Squadron's activities on 9 July are fairly typical of the Battle of the Somme. Before dawn, the pilots of 'A' Flight were out on the field with the groundcrew warming up the engines for the early sorties, the familiar sound of rotary engines blending with the distant clatter of the trenches to break the early morning stillness. At 03.50hr two Moranes took off into the pale eastern daybreak — one to complete a tactical and flash reconnaissance while its escort dropped two bombs on a likely looking railway siding. Before they landed, a Nieuport was up looking fruitlessly for the enemy dawn patrol, and at the same time two further Moranes took some photographs of the enemy lines. After breakfast a further Morane got airborne with an infantry officer in the observer's seat: he wanted to orientate the exact line of the enemy's forward trenches for his own company's coming assault. All morning the aircraft were coming and going. At lunch time, the Nieuports were off together on an offensive patrol, but few enemy aircraft were seen, and none engaged. Later still, the Moranes flew a patrol from Steenstraat to Wytschaete, and so it went on. Finally, at 21.41hr, the last aircraft settled to earth on Bailleul airfield in the gathering gloom. Exhausted aircrew gathered in the Mess for supper and a beer, the last servicing tasks were completed among the bodies of sleeping groundcrew, and, in the distance of the starlit summer night, the clatter of the trenches went on.

After a month's heavy fighting on the Somme front some of the fire went out of the great Allied effort, and the RFC squadrons were able to concentrate once more on offensive action in their own areas. On 2 August three of No 1 Squadron's biplanes escorted a formation of No 2 Brigade aircraft which bombed the Zeppelin sheds at Brussels. This involved a radius of action of 70 miles, the Squadron's deepest penetration of enemy airspace throughout the war. In spite of the two hours spent in hostile skies, only one German aircraft was seen, and it made off quickly at the sight of the escort. The Nieuports, meanwhile, were being used for shorter offensive patrols and had more of their share of the fighting. Lt B. J. Moore, flying a high patrol at 17,000ft, spotted an Albatros two-seater circling between Armentieres and Lille. Diving out of the sun, he riddled the enemy machine with a well-placed burst. Immediately, a long shaft of flame burst from his quarry's fuselage, and although Moore broke hard out of his attack, he was close enough to feel the heat of the burning aircraft on his face. As he returned for a further attack, he saw that the observer had fallen out. Without a parachute, his victim's end was inevitable, and he plunged to his death alongside his pilot's flaming coffin near Frelingham.

In early July, two of the new two-seater Nieuports had been added to the Squadron's inventory. With four types now being flown on the same unit, pilots tended to concentrate their flying on just one, and aircraft were individually allocated according to experience and personal preference where possible. A number of the pilots, including Latta and Moore, had been applying their aggressive temperaments to good effect in the single-seat cockpits of the Nieuport Scouts for some months and they were bitterly disappointed when, in mid-August, the Scouts were withdrawn and given to another unit. Unfortunately this deprived the Squadron of its integral specialised escort flight. The Scout pilots were naturally disgruntled at having to pair up with observers once more and apply themselves to the more mundane, but no less dangerous, tasks of spotting and reconnaissance.

Throughout the early autumn of 1916 the pace continued unabated. The Squadron's earlier trials with Brock flares had given rise to a standard system of contact patrols, whose communications with artillery commanders prevented any shelling of Allied troops. Throughout August and early September the Squadron flew contact for the 41st Division, I Canadian Corps and the 4th Canadian Division. Opportunity bombing during offensive patrols once more began to play an important part in operations, and in October a number of impromptu raids were made on hutments, dumps and railway installations. These attacks were usually carried out in flight strength, the six or eight aircraft involved all being armed with four 20lb bombs.

The 'Fokker menace' had passed; the predominance of this German monoplane had diminished with the development of new defensive tactics. But during September 1916, new types of German scout began to appear over the front and show an alarming turn of speed and manoeuvrability. Later identified as the Fokker DI, the Halberstadt and notably the Albatros DII, their early successes resulted in serious rethinking at RFC Headquarters. This led to the decision to form additional specialised fighting scout squadrons equipped with single-seat machines. Imagine the delight of No 1 Squadron's pilots when they learned that they were shortly to be re-equipped entirely with Nieuport 17 Scouts. It appeared that their earlier record with the 'Baby' Nieuports had found considerable favour in Trenchard's eyes, and that the Squadron's selection for this envied role was the personal decision of 'Boom' himself. The news was received with dismay by the observers, who had long feared such a move, but the pilots were naturally overjoyed. The long hours of artillery patrols were almost over, and the possibility of revenge was in sight for the many months of droning through the Archie as little more than 'Fokker Fodder'.

The decision that decided No 1 Squadron's future took some time to gain effect, and army co-operation work continued into the new year. Another Christmas arrived and brought with it Maj G. C. St P. De Dombasle to take command of the Squadron. The new CO relieved George Pretyman on Christmas Eve, and on 29 December he supervised the transfer of the unit from the 2nd (Corps) Wing to the 11th (Army) Wing of No 2 Brigade.

The officers of No 1 Squadron in December 1916, shortly before the departure of the Observers. Maj Guy de Dombasle and Maj George Pretyman are seated in the centre of the second row. Capt B. L. Moore is seated third from the right.
Squadron Archives

CHAPTER THREE

No 1 (Fighter)

No 1 Squadron greeted the New Year of 1917 with the arrival of six brand-new Nieuport Scouts. Training for the new role started immediately and all pilots were soon converted to type as the full complement of machines arrived. It fell to the remaining Moranes to conduct the Squadron's operational flying during January, and artillery co-operation was continued until the 29th, when the last two-seat aircraft to be flown in action by the unit was struck from the inventory.

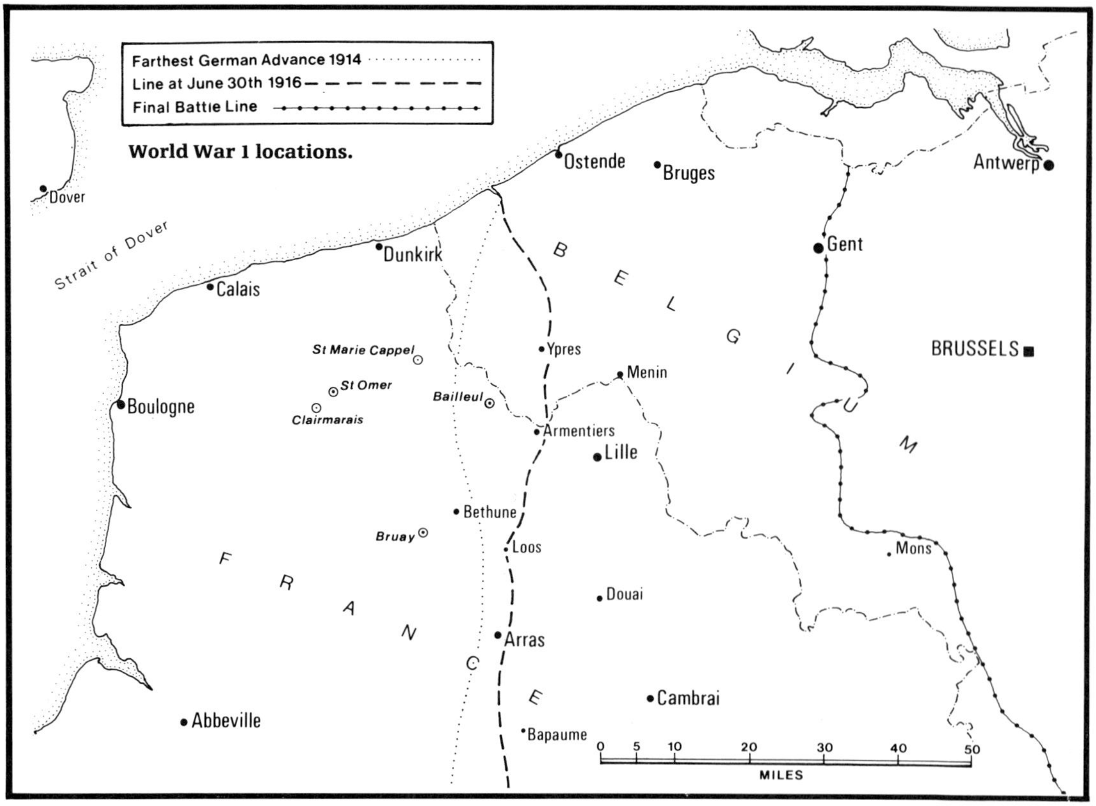

At the beginning of February No 1 was declared operational as a scout squadron and assumed the title of No 1 (Fighter) Squadron, Royal Flying Corps. Although the system of adding the role to a squadron's title is no longer used, No 1 still carries the (Fighter) suffix unofficially to this day, as a mark of pride in its outstanding record in the role to which it has been assigned (with one brief exception), for almost 70 years.

The first Squadron patrols with the new aircraft were carried out on a line from Bailleul to Abeele on 9 February. The first few days were uneventful, but on the 15th a formation ran into enemy aircraft, one of which was shot down by Lt V. H. Collins, and another was driven down by the rest of the patrol. Over the following weeks, increased enemy air activity produced many engagements and several additions to the Squadron's growing score-board. It was noticeable, however, that the enemy's fighter tactics were improving, and kills were becoming increasingly difficult to achieve. A significant factor here was the recent introduction of the *Kette*, a loose formation of four German scouts comprising two mutually self-supporting pairs, whose combined lookout made it difficult for British pilots to sneak up from behind.

At this stage of the war, only a few personalities of the *Luftstreitkrafte* — the Imperial German Air Service — were known to the pilots of the RFC Squadrons. Max Immelmann was renowned for the half-looping manoeuvre he had developed in early combats, and the name of Oswald Boelcke, the late commander of *Jagdstaffel 2*, had been often heard in British crewrooms during his short career. Both these men were now dead, and only the individual markings of the Fokker and Albatros scouts served to identify the adversaries that were met daily high above the trenches.

But as the days lengthened, reports grew more frequent of a new elite squadron of enemy scouts. The leader flew an Albatros of a new design which was finished in a menacing blood-red colour scheme. The machines he led were similarly painted and their pilots were identifiable by gaudily coloured rudders and fuselage stripes. As the slow British two-seaters fell before their guns in increasing numbers, Intelligence was to learn that this was the newly formed *Jagdstaffel 11*. Their base was at Douai, just 30 miles down the trenches from Bailleul; their commander was Baron Manfred von Richthofen.

Not all German squadrons, however, were as formidable as *Jasta 11*, because it was now enemy policy to concentrate talented pilots into elite squadrons. Remaining units, denuded of their best young leaders, were thus less than a match for the British fighter squadrons, despite the rapid introduction during February and March of the new Albatros DIII, recently seen in the hands of von Richthofen. Powered by a 200hp Benz, the new Albatros was not much faster than the DII, but its rate of climb was superior to that of its predecessor or any of its adversaries. More important still, its reserves of power enabled it to maintain height in the sustained turns of a combat, which would place it at an advantage after the initial manoeuvres of an engagement. Its sleek, fish-like fuselage and raked biplane wings gave it a performance margin over the Nieuport that No 1 was to find painful before the year was out.

In spite of all this, the first engagement with the DIII was very successful. Capt C. J. Q. Brand, who had recently been posted to No 1 as a Flight Commander, was leading a patrol of three Scouts in the Zillebeke area, where an LVG reconnaissance aircraft was seen and attacked. Brand pumped three-quarters of a drum of Lewis ammunition into the German aircraft, which burst into flames. As it fell, the wings were torn off and fluttered down near the fuselage, which crashed at Zandevorde. An Albatros DII that was coming to its aid was also attacked and shot down by Lt C. C. Clark. On the way home, two further LVGs were engaged near Polygon Wood, but before they could be destroyed three of the new DIIIs attacked the patrol from up-sun. In the sharp and vicious combat that followed, Christopher Brand scored again by driving one down out of control. The other two made off. Two further DIIIs were met over the Ypres-Comines canal and one of these was likewise driven down, this time by the whole formation. For this action, Capt Brand received a personal commendation from Trenchard.

To the south, the German army was withdrawing strategically from the Somme to the prepared defences of the Hindenburg line. As a fighting scout squadron, No 1 still had a reconnaissance role, and in late March carried out valuable photographic sorties of the area behind Messines and Wytschaete. Many escort missions were also being flown, accompanying the ungainly FE2s and BE2cs deep behind enemy lines on bombing and long reconnaissance sorties.

While the Germans strove for air supremacy,

the RFC was in an unhealthy shape to meet or mount a major offensive. Of the 900 aircraft allocated to operational units, only some 550 were really combat fit. Of the rest, too many were either obsolescent or, in the case of the newly-arrived SE5s and Bristol Fighters, completely untried in action. It thus fell to the few operational fighter squadrons to carry the brunt of the offensive air battle against men like von Richthofen and Voss. Such was the state of the RFC's equipment in the spring of 1917 that Trenchard was forced to report: '. . . you are asking me to fight the battle this year with the same machines as I fought it last year. We shall be hopelessly outclassed'. Certainly the outlook was grim, especially for the crews of the two-seat bomber and recce aircraft that flew, ageing, heavy with fuel and lightly escorted in the face of the proud new Albatros.

There now began another great Allied offensive centred on Arras, some 25 miles south of Bailleul and just four miles from Richthofen's base at Douai. Conceived by the suave French General Nivelle, the plan was for British forces to attack on a 25-mile front from Arras to St Quentin, while the French were to strike on the Aisne sector. Sir Douglas Haig was unhappy with this strategy, feeling that he had had the plan foisted on to him by his political masters. Nivelle was, he considered, a rascally charlatan. In such an atmosphere of Allied mistrust, it is hardly surprising that the offensive was to prove a grisly and expensive failure. To make matters worse for the RFC, its great Commander, who so despised human ills and weakness, fell sick with German measles on All Fools Day. 'Bloody April' had begun.

The battle opened for No 1 on 5 April, when four squadrons of scouts were detailed to destroy all enemy kite-balloons on the Arras front. A new weapon developed for this task was the Buckingham incendiary bullet, which was sufficiently flat-nosed to tear a great hole in the canopy of the balloon. They found that it was not particularly successful, because the few balloons that were hit were hauled to safety as they deflated, and none was claimed as destroyed that day. To add to the pilots' misgivings, a rumour was rife that the Germans would look upon the Buckingham as a dum-dum, and that captured pilots found carrying it would be shot. When the Squadron was sent out three days later, the crews therefore armed themselves with the more familiar Le Prieur *Torpeille*. Pressing home its attacks with customary verve, the Squadron destroyed two balloons, claimed by Capt Brand and Lt E. S. T. Cole, but suffered the loss of Lt R. J. Bevington, who was later reported to be a prisoner.

Lt Stuart Keith-Jopp drew this sketch of his own first combat. In his diary he wrote: 'As he passed underneath, I saw it was a two seater, with one red and one yellow wing. "A Hun" thought I, "a real Hun — What do I do now? — Attack I suppose." So I cartwheeled and went down vertically at him. On getting close I had a peep through my Aldis sight, but it had blurred with moisture, so I just pulled the trigger and howled with joy, as through a sort of red mist, I watched the tracers going round him. The Hun flew straight on without a sign, till I was about 10 yards from him, and then he dived vertically too. I followed, and missed his rudder by about 2 feet in my effort to keep the spray of tracers on him . . . then my machine got over the vertical and began to spin on its back.' I got one glimpse of the Hun going down, and was then chucked out of the seat into my belt. Everything fell out except me, and the Huns got a present of my automatic and Very pistol. Round and down we went, faster and faster, with me making frantic efforts to pull the joystick back, but always being thrown on to it and knocking it forward again. Finally I got the rudder straight and pulled out at 8,000 feet. My ears felt pretty rotten after falling a mile, so I started for our side, getting thoroughly plastered by Archie on the way. I looked for my Hun, but he had gone. I wish I knew what had happened to him'. *Squadron Archives*

As in the previous spring, the weather during April was very bad. The task of escorting the two-seaters on their bombing and reconnaissance missions was rendered virtually impossible by low cloud and poor visibility. The slow aircraft, forced by the weather to cross the front lines at low level, now fell prey to small arms fire and were hacked out of the sky in droves.

Because of the weather, the Squadron flew a

large number of offensive patrols during the battle, scouting ahead of the heavier aircraft to try to prevent the enemy fighters falling on an easy target. One such patrol of nine Nieuports was sent off on the 22nd, but while over enemy territory one of the aircraft signalled that it had engine trouble. Lt E. S. T. Cole was waved off by his leader to escort the lame duck to safety, and on the way home he spotted an enemy balloon flying at 2,000ft near Wervicq. After ensuring that his charge was safe he returned to the attack, but on his approach the balloon commenced a rapid descent. In spite of heavy Archie he pressed the attack home, and his quarry burst into flames when only 20ft from the ground. As Cole was pulling out, however, his engine was hit by ground fire and started running very roughly. He made for friendly lines as fast as he could with the engine cutting intermittently, which caused him actually to touch the ground more than once before scraping home at less than 50ft to force-land on the Allied side of the trenches. Lt A. W. Wood was not so lucky, having been shot down and taken prisoner during the same patrol.

The following day Lt S. McKercher was wounded during a patrol, and on the 25th, 2nd Lt Collins and Lt L. J. Mars collided over Bailleul and spun down locked together. They both survived with mild concussion.

Although a further Nieuport was lost on the 26th (Lt A. Burberry being taken prisoner), No 1 had got through 'Bloody April' very lightly: analysis of the casualty figures of that dreadful month show that an RFC pilot flying on the Arras front for four hours a day — a fairly normal figure — could expect to live through just 23 days.

With the arrival of May the weather improved, and the losses of the two-seater squadrons fell dramatically. The clearer skies also meant that contact with the German scouts became more frequent. The Squadron lost a valuable pilot on 1 May when Capt Brand was wounded in an encounter with an Albatros. In his time with No 1 he had accounted for six enemy aircraft and several balloons; he later went on to become one of the most successful pilots of the war.

The recent casualties suffered by No 1, although light by some standards, needed replacement. Most of the new pilots received during April and May quickly became useful members of the Squadron and, in a couple of cases, really exceptional fighter pilots. 2nd Lt W. C. Campbell stood out immediately. At 33 he was far older than most new arrivals. Born of a Scottish father and French mother, Campbell was a wine importer by profession, with a taciturn manner that initially set him apart from his colleagues. Lt Philip Fullard could not have contrasted more vividly with Campbell. His extrovert personality disguised the hard core of the real professional and when not joining in the fun at a Mess party, Fullard could be found down in the hangars polishing and servicing his guns. On non-operational flights he would strip his guns on to his lap and practice airborne stoppage drills until he could do them with his eyes shut. He was, as a result, never to have a stoppage during the 5,000 rounds he fired in combat that year. Everything in his academic and sporting life had borne the same professional stamp which marked him as a young man to watch.

18 May was a bad day; Lts Lindsey-Drummond and M. G. Cole were lost when strafing balloons and Lt T. H. Lines was shot down and taken prisoner. Lt Stuart Keith-Jopp recorded in his diary that night that the loss of Lt Lines, who went down during a strafing attack on a balloon, upset Campbell very deeply: 'At dinner Campbell cursed and damned, walking all round the table until we turned him out and had more drinks. He continued outside, and the next morning went out and shot down a balloon in revenge against orders.' Lines had arrived with Campbell, and had been his best friend.

Capt W. C. Campbell DSO, MC and Bar, who destroyed 26 aircraft and 10 balloons between May and October 1917. *Squadron Archives*

In addition to the 'off pats', many line patrols were being flown at dawn and dusk. The morning flights were very unpopular because the dazzling sun would fill the eastern sky where any amount of Huns could lie in wait. The morning mist would then rise to obscure Bailleul and the Nieuports would have to divert to any convenient cornfield until it dispersed. In the evenings, however, the boot was on the other foot, and although they did not always produce a successful engagement, the dusk patrols were more popular. Let Keith-Jopp describe one flown on the evening of 25 May:

'I like evening patrols. The fat sun is then behind you, and you feel a proper hired assassin and ready to engage in any dogfight going. The cloud formations are usually very fine, as with the setting sun, the shafts of light strike across and up. So whilst you, at 18,000ft, are flying in brilliant sunshine, the earth is blue with the dusk and the air beneath is full of shafts of red and golden light. This evening we did a patrol beyond Menin. Suddenly, our leader went into a spiral dive. We all followed. I was in the rear, and it was beautiful watching the other five machines cascading down. Beneath us was one of these shafts of light with the blue ground beneath. After going down about 6,000ft, I saw several shadowy machines against this blue, and then suddenly a bright red machine dived through the shaft of light. For a second or two, it shone like a red flame and disappeared into the blue beyond.

Lt Stuart Keith-Jopp (arm in sling), at Buckingham Palace after losing his hand at Bailleul in May 1917.
Squadron Archives

This was probably Richthofen and his circus, as they came up into our sector of the line that day. We followed them down some way, and then gave up.'

The following day the Squadron was involved in a full scale battle with a large formation of enemy scouts, having been bounced just across the lines while escorting six FE2bs on a reconnaissance mission. The pilots gave a good account of themselves, claiming three Albatros destroyed. Lt J. C. C. Piggott, rejoining the fight after a breath of fresh air, had the bad luck to be hit in the lung by a stray bullet. Keith-Jopp was in the thick of it and fought off repeated attacks by a couple of Albatros V-Strutters that fastened on to his tail, riddled his engine and fuel tanks, and put a bullet through the hem of his coat. He managed to glide the remaining five miles to Bailleul, where a visiting general berated him for leaving his patrol! Later that day Lt R. R. Mackintosh was lost on his first patrol. This annoyed Keith-Jopp: 'I'm fed up. My beautiful machine has been lost. A new pilot stole it today, and has not come back. It is too sickening'.

The next day the Squadron recorded the loss of Capt J. R. Anthony, who unlike Mackintosh

A camouflaged Nieuport 27 of No 1 Squadron, visiting a squadron of SE5s. This photograph was taken by Capt James McCudden VC.
Squadron Archives

was not listed as captured. His place as Flight Commander was taken by Capt T. F. Hazell. During the same patrol Capt Honnet, who was leading, fainted at 18,000ft and spun down to low level before he recovered. In the days before aviation medicine, the lack of oxygen was considered more of a little-understood nuisance than a mortal danger.

On the night of 31 May enemy artillery scored a direct hit on an ammunition dump in nearby Kemmel, causing some consternation in the Mess and blowing all the windows out. Campbell was playing diabolo at the time. As the other pilots crowded out from under the tables he casually remarked, 'Good for ze Hun', and just carried on playing.

The following morning the Squadron lost another valuable man when a Very pistol exploded in Keith-Jopp's hand and he lost an arm. He later became a flying instructor in England and flew in the Air Transport Auxiliary in World War 2, delivering over 1,300 fighters. As 2nd Lt T. Perkins was killed in a flying accident on the same day the record for May looks pretty bleak, except that 35 enemy aircraft were claimed during the month, of which no less than 10 were destroyed by Lt Philip Fullard.

Another tremendous explosion rocked Bailleul on 5 June when one of the rare enemy daylight bombing raids found its mark on an ammunition train drawn up in the airfield siding. Heavy casualties were suffered by the railway staff, and the smoke from exploding shells closed the airfield for over three hours.

The Battle of Arras had petered out by 15 May and in early June the offensive was switched to the Messines-Wytschaete ridge. Reconnaissance of this area had been completed by the Squadron during March, and the significance of its task now became apparent. As the battle opened, No 1, in addition to its patrol work, was tasked with numerous close air support sorties, attacking troops and transport with guns and 20lb Cooper bombs. On 8 June the pilots were attacking the enemy in this fashion in the Poelcappel-Wervicq area, when they had the misfortune to run full tilt into a strong patrol from *Jasta 27*. Although two enemy scouts were destroyed, 2nd Lt R. S. L. Boote was seen to crash, and as they fought their way home they noticed that Lt J. Slee was also missing. They were later relieved to hear that both their comrades had been captured alive. Slee had been brought down by the enemy Squadron Commander, whose name was Hermann Goering.

The following day they were again engaged by *Jasta 27* aircraft, but managed to shoot down three of them for the loss of Lts F. Sharp and W. J. Mussared, both of whom were taken prisoner. Mussared had lied about his age to join the RFC, and had shot down eight enemy aircraft before spending his 17th birthday behind bars.

Good targets were now plentiful, and ammunition was often exhausted on these close support sorties. On one occasion Campbell, having attacked and driven down two balloons, engaged and dispersed troops to the east of Polygon Wood. Returning to base, he saw more troops entraining at Houthem siding and attacked them with customary gusto. Running short of ammunition, he fired all his Very pistol cartridges at them, scattering them wildly in all directions. Campbell, like Fullard, was beginning to make quite a name for himself. Awarded his first MC for the action just described, he threw himself into every engagement with complete disregard for his own life. On 16 June, in a combat with a large number of enemy aircraft, he destroyed two Albatros two-seaters close to Bailleul before running out of ammunition. After dashing home for more he managed to complete his hat-trick within 40min.

The time had now come for Maj Guy de Dombasle to relinquish command of No 1 Squadron. On 20 June he handed over to Maj A. Barton-Adams, who took command at the height of some of the heaviest fighting that the Squadron had yet experienced. No 1 was flying all day and every day. Little time was available now for training and settling in new arrivals, most of whom were coming to the front with less than 30 hours' flying experience. It is amazing that the Squadron's casualties were not heavier in view of the determined opposition and the fatigue that was bound to result from months of intensive dogfighting. Yet in spite of the losses of Lts C. Street and T. Littler on 5 July, the Squadron's balance sheet was beginning to look quite healthy, as nearly 100 victories had been logged since the introduction of the Nieuport Scout.

In the early part of the war, enemy airfields had not been regarded as suitable targets for the gentlemen aviator, but the cold facts of combat were now beginning to over-rule chivalry. On the 31st, orders were received to attack the airfields at Herseaux and Morveaux in support of the opening offensive of the Third Battle of Ypres. An Albatros was set on fire at Herseaux and mechanics were scattered in all directions, whilst at Morveaux one of the pilots swept in low and fired 300 rounds at a dozen two-seaters lined up on the ground. He was followed by Lt Campbell, who raced in at 50ft to drop his 20lb Coopers among a cluster of six Albatros scouts, three of which went up in flames. During his first pass Campbell was wounded by a gun team alerted by the passage of the leader and, by now fighting mad, he Immelmann-turned at the edge of the field, dived back and wiped out the gun-crew. Pulling out from this attack he expended his remaining ammunition on the headquarters building, and made for home. His wounds were not severe, but he was moved from the Officers' Mess into the new hospital set up in the asylum. After a well-earned rest and a spell of leave he returned to the fight by the end of August, sporting a Bar to the MC that he had been awarded the previous month.

Campbell was not the only pilot to excel on 31 July. The others had had a field day in the absence of enemy scouts, strafing a column of enemy transport on the Gheluwe-Menin road and attacking a battalion of infantry at Tenbrielen. They had achieved complete surprise in the latter case by diving out of a rainstorm. These attacks were led by the newly-promoted Capt Fullard, who was, like Campbell, later awarded his second MC for his part in this successful day's operations.

By this time, patrol engagements by opposing scouts were frequently building up into battles fought out by 50 or 60 machines on each side. At Douai the successes of *Jasta 11* had led the German High Command to promote von Richthofen to the cavalry rank of

Capt W. V. T. Rooper, commanding 'B' Flight, was killed in action on 9 October 1917.
Squadron Archives

Rittmeister to form *Jagdgeschwader I*, a big wing of 48 aircraft comprising the successful *Staffeln 4, 6, 10* and *11*. The Allied reaction was to encourage squadrons to patrol in company and to inform the airfields of news of any battle building up. Any airborne scouts seeing the dogfight develop would climb at full speed to join in at the top. On 16 August, Capt Fullard was leading 'A' Flight on an offensive patrol when a 'beehive' was spotted some distance away — *Jagdgeschwader I* was dancing like flecks of blood with numerous Allied machines. Joining the battle he saw a Spad attacking two enemy scouts and a further Albatros turning in on the Spad's tail. Seizing his opportunity, Fullard snapped in a quick burst at the latter machine, which fell in flames and was seen to crash. Immediately he was engaged in a duel with another red Albatros, which he also destroyed. With fuel and ammunition running low, he collected his Flight and made for base with 2nd Lt J. H. C. Nixon nursing a Parabellum wound in his leg.

Three days later 2nd Lt H. E. A. Waring was shot down in a similar fight, but survived to join the 19 members of No 1 Squadron already in the prison camps. One of these, Lt W. C. Smith, had been wounded before his capture on 13 July, and was repatriated from hospital the following year.

In early September the Squadron was re-equipped with a later model of Nieuport, the Type 27. The earlier Nieuports had proved somewhat tail-heavy and unstable below 8,000ft and were therefore not really suitable for ground attack. With a strengthened and altered tail and a little more speed, the Type 27 was more suited to strafing.

After some intensive ground attack operations in late September, No 1 was one of several squadrons engaged over Passchendaele Ridge between 5 and 9 October. Providing top cover for bombing raids produced five days of intensive combat, and the scoreboard continued to rise in a most satisfactory way. Patrols flown on the 9th engaged a total of 30 enemy aircraft, of which six were brought down. A notable milestone was thus reached. The following extract from the official RFC communique — known to all as 'Comic Cuts' — tells the story:

'The enemy machines brought down today by No 1 Squadron bring the Squadron's total of EA shot down or driven down out of control since commencing work as a scout squadron on 15 February 1917, to two hundred machines'.

Capt Philip Fullard DSO, MC and Bar — the Squadron's highest scoring fighter pilot with 42 aircraft and three balloons destroyed in eight months. He broke his leg playing football in December 1917. *Squadron Archives*

Two hundred up and a lull in the fighting gave time for a brief celebration and a little relaxation from 10 long months of heavy fighting from Bailleul. Even then the shadow of those they had lost served to temper pride with sadness, for on the mission that had made the Squadron one of the top-scoring Allied units of 1917, Capt W. V. T. Rooper had been killed and 2nd Lt M. A. Peacock was taken prisoner. Rooper had led a Flight throughout the recent action with considerable success, and some mystery surrounded the discovery of his naked body at Passchendaele, over 15 miles from the scene of his final dogfight.

In the middle of October the weather set in to prevent any effective offensive patrol work. The various actions of the Third Battle of Ypres were still grumbling on and the Germans now took a leaf out of the RFC's book by sending aircraft out to raid Allied trenches. This was for the pure nuisance value, but the morale of the troops dictated that they should be driven off. These sorties were not much fun,

Right:
No 1 Squadron in the snow at Bailleul during January 1918. The aircraft are a mixture of Nieuport 17 and 27s: their serial numbers have been erased by the wartime censor.
Squadron Archives

Bottom right:
An SE5A pilot of No 1 Squadron shows off his Lewis gun on its Foster mounting that enabled the pilot to reload in flight.
Squadron Archives

conducted as they were under a low cloud base in the midst of a heavy artillery barrage. Capt G. A. H. Pidcock, attached to the Squadron from No 60 for a Fighting Instructors Course, reported that he was not at all fond of these sorties:

'Being a visitor to the Squadron, I was always being given the nastier jobs. We were usually called up too late to drive these Huns off our trenches, and the job was made very dangerous by the extremely heavy barrage. The air was often literally full of shells'.

Apart from occasional reconnaissance sorties and the interception of sneak raids against Allied trenches, the pilots and groundcrew were now able to rest awhile, overhaul the aircraft and consolidate training in the air when the weather permitted. Campbell was posted and the Squadron was delighted to hear that both he and Fullard had been awarded the DSO. Campbell's score on his departure stood at 26 aircraft and 10 balloons destroyed.

The drizzle of October gave way to the fogs of November and the Battle of Cambrai, which opened on the 17th. In this action the Squadron was detailed for the dangerous task of low level support work in appalling visibility. Although this meant that the heavy ground fire would claim fewer victims, there was the ever-present danger of flying into trees and other obstacles during attacks. Surprisingly, only one aircraft was lost during the battle, when 2nd Lt C. E. Ogden crashed in fog behind enemy lines to become a prisoner on 5 December. However, Lt W. W. Rogers managed to bring down two large Gothas on 12 December, one of which crashed in flames.

Second Lieutenant R. O. Phillips was wounded on 18 December, and on the same day Philip Fullard broke his leg when leading the Squadron football team. The invaliding home of its most successful pilot was an immeasurable loss to No 1 Squadron. An outstanding Flight Commander and patrol leader at a very tender age, Fullard had been a constant source of inspiration to the more recent arrivals. Armed with a DSO, MC and Bar, he returned home in a blaze of publicity with 42 aircraft and three balloons to his credit, all scored during his six short months with No 1 Squadron. Fullard's scoring rate stands as one of the Squadron's outstanding wartime achievements: in six months he had become the ninth ranking British fighter pilot of the Great War. He stayed on to serve in the RAF after the war and retired in 1947 as an Air Commodore, having commanded No 246 Group during World War 2.

In the opening month of 1918, the offensive patrols began once more. Capt Tom Hazell, another of the Squadron's outstanding Flight Commanders, added a further five enemy aircraft to his score before being posted in February to No 24 Squadron with a total of 19 kills during his time with No 1, despite being wounded in a duel with von Richthofen earlier in the year. He was to go on to become another of the top scorers of the war with 41 victories.

By this time the Germans were beginning to introduce the Fokker DrI Triplane. It soon became clear that this manoeuvrable little fighter would completely outclass the Nieuport. No 1 Squadron thus learned with relief that it was to be re-equipped immediately with the SE5A, the latest version of the Farnborough-designed scout that had first seen service during the previous April with No 56 Squadron.

At the end of January the pilots were pulled out of line to re-equip. They found their new mounts to be remarkable machines. Designed by H. P. Folland at the Royal Aircraft Factory, the Scouting Experimental No 5 was powered by the 200hp water-cooled Wolseley Viper, could climb to 15,000ft in under 23min and fly there at 117mph, well below its service ceiling. Its broad wings embodied the latest aerodynamic features, and enabled it to turn as well as many lighter scouts, but at the same time gave sufficient stability to provide an excellent gun platform. Mounted on the port side of the engine cowling was a Vickers gun, synchronised to fire through the airscrew using the Constantinesco hydraulic interrupter device, and on the centre section of the upper wing lay a Lewis gun on the Foster mounting that enabled the pilot to haul it down for reloading. After a short conversion course and a period of gunnery training, the Squadron commenced line patrols that were run concurrently with further combat training on the new aircraft.

Trade during the latter part of February was fairly light, and gave rise to the belief that the Allies were beginning to get the upper hand. But if the enemy was down he was far from out, for during the opening weeks of March increased German air activity and troop movements warned that something big was afoot. On 21 March the Germans launched their massive spring offensive on a broad front. After a few days of intensive bombardment their forces broke through the lines in several places, and the Allies were faced with the prospect of mobile warfare for the first time in three years. Every available unit of the RFC was thrown into the battle to harass the advancing enemy with bombs and machine-gun fire. No 1's pilots were in the air from daybreak until after sunset.

From dawn on 26 March until dusk on 3 April, No 1 Squadron was detached on a daily basis to reinforce the 10th (Army) Wing, which was heavily engaged with advancing German infantry in a serious breakthrough to the south of Arras, on what became known as the Bapaume front. The SE5s were flown south to Bruay each morning and returned, battle-scarred and weary, to their own airfield at night.

On the afternoon of the first day at Bruay, the Squadron, armed for the first time with 112lb bombs, attacked the enemy trench lines to harass units moving out through no-man's land. Spearhead units of the advance were strafed as they fought their way into Allied territory. Three aircraft were lost in the first day's action, 2nd Lt Denovan being posted missing while Lt A. Hollis and 2nd Lt W. M. R. Gray were taken prisoner by the advancing enemy. In the afternoon of the following day an enemy bivouac was bombed by five of the Squadron's aircraft. Troops, horses and trans-

port were dispersed in absolute disorder, and while they stampeded the pilots circled round making firing passes, and left the whole camp in flames. Later in the day columns of infantry were bombed, while transport that was attacked in the vicinity of Croiselles was described by one pilot to 'pile up in a heap'. During this last patrol of the day, an interfering Albatros was engaged at point blank range and shot down near Bapaume.

When the pilots returned to base on the evening of the 28th after a further day's heavy fighting, they learned that the German advance was getting uncomfortably close to Bailleul, and that they would be pulling out in the morning. The CO and Lt Gibbons, the Recording Officer, were furiously packing up the headquarters; the groundcrew under Capt Noel were dismantling and loading their equipment ready for the hurried move and the officers spent their last night in the shadow of the asylum collecting their few personal belongings and despatching them with the batmen to their new base at St Marie Cappel. At daybreak the SE5s once more took off for Bruay, and the ground party pulled out for its new airfield alongside Highland troops fighting a desperate defensive action against the now jubilant Germans. The pilots flew into St Marie Cappel that night too tired to care much about their new surroundings and snatched a few hours sleep before returning to Bruay at dawn.

This was to be the pattern of the next five days. Each pilot flew five, sometimes six hours of bombing and strafing sorties per day, finishing with a ferry flight and often a night landing. Fatigue was setting in, but there was to be no immediate respite.

Meanwhile, in England, the final arrangements were in progress for a historic change in the structure of British military aviation. The need had been recognised — and also hotly opposed — for the establishment of an independent air arm.

On 1 April 1918 the Royal Air Force came into being. However, the honour of becoming the senior squadron of the first independent air force was rather lost on a group of very tired men in Flanders. For some time they would continue to use a miscellany of military and naval ranks, and wear the uniforms and

Bailleul after the artillery barrage which started on 30 March 1918, the day after the Squadron pulled back to St Marie Cappel. *Squadron Archives*

badges of the regiments from which they had been seconded.

The pressure on the Bapaume front eased slightly by 4 April. Because of an abundance of targets nearer home, no more operations were flown from Bruay. Although the emphasis was still on ground attack for the remainder of April, many desperate engagements with enemy aircraft were recorded. The most notable of these occurred on the 7th, when Capt G. B. Moore led 'B' Flight full tilt into a superior number of red Fokker Triplanes from *Jagdgeschwader I.* Taking on the 'circus' at adverse odds was always a very bold move, and it was at once apparent that they had a hard fight on their hands. The initial engagements took place at 10,000ft, and in the whirling dogfight that followed both patrols soon fought their way down to low level. The British pilots had immediately noticed the white tailplane of the leading enemy scout — von Richthofen himself was in the fray.

With no score on either side, the little cloud of twisting fighters had descended to 2,000ft. Suddenly an extraordinary thing happened. Moore, flying No C1083, was turning to attack one of the Fokkers when he exploded in a great fireball. The fight ceased as suddenly as it had begun, and the leaderless patrol found itself alone in the sky. Back at base they were at a loss to explain Moore's death. No one had seen a Triplane on his tail at the time, and the explosion seemed too great to have been caused by a burning fuel tank. The Squadron's Recording Officer therefore reported his loss as being due probably to a stray artillery shell from the nearby front line. The pilots involved were later annoyed to hear that Moore had been claimed as Manfred von Richthofen's 77th victory. The matter was never satisfactorily resolved and the Squadron had lost yet another of its more seasoned pilots. Moore, a Canadian, had destroyed 12 enemy aircraft before his mysterious death.

The fortunes of No 1 during the latter part of the German advance, by now known as the Battle of Lys, were mixed. It suffered badly on 10 April, when Lts D. M. Bissett, F. P. Magoun and H. W. McKeague were all wounded by ground fire, while Lts W. D. Patrick and G. A. Mercer were brought down to spend the rest of the war in captivity. For Magoun there was some consolation in the award of the MC for his part in the operations of recent weeks. Another recipient was 2nd Lt P. J. Clayson, a recent arrival who had pressed home attacks with great determination on his early sorties. The weather on the following day was dismal, but the cloud base lifted sufficiently in the afternoon for Capt W. L. Harrison to lead three aircraft on a low level offensive patrol. To the south of Armentieres they spotted four of the new Pfalz DXIIs and wheeled into the attack at a scant 400ft. Almost immediately their four intended victims were reinforced by five comrades, while a Sopwith Camel patrol from No 54 Squadron joined in. Harrison saw an enemy scout on the tail of one of his wingmen and fired 75 rounds into it to send it spinning into the ground on its back. Lt F. R. Knapp, Harrison's deputy leader, had meanwhile tangled with the leader of the Pfalz patrol and nailed him with two quick bursts. His victory had to be confirmed by the leader of the Camel patrol as Knapp was too busy fighting off four further enemy aircraft to watch his prey hit the ground. Knapp was shot down and captured the following week.

On 12 April, patrols from the Squadron ranged far and wide on a day of rain-washed sunlit skies and pressed home attacks all day long on targets around Estaires, Merville, Neuf Berquin and Steenwerck. A patrol of six aircraft strafed infantry 3,000 strong to the west of Steenwerck in the face of massed small arms fire through which they flew unscathed. Another section inflicted heavy damage on a column of transport at Merville. The squadron fired 5,000 rounds and dropped 125 25lb bombs that day, out of a total of 279 dropped by all four squadrons of the Wing.

The following day the German advance rendered St Marie Cappel as untenable as Bailleul, and once more the Squadron was forced to move back, this time a further eight miles to Clairmarais, near St Omer. The loss of their intermediate base meant little to the weary pilots. For just a fortnight it had been their home, an unknown field on which to refuel and take on ammunition, a brief sanctuary to rest a weary limb, a place they hardly knew by daylight.

Capt C. C. Clark and Lt K. C. Mills, who usually flew and fought as a pair, did well in April, sharing an enemy scout at low level over Lompret on the 21st and claiming a further three between them the following week. Clark, flying alone on the evening of the 23rd, took on eight Fokker triplanes single handed and shot one down before escaping in a cloud bank. In the low cloud and mists of the 24th, the Squadron flew bombing sorties all day but encountered no enemy aircraft at all: it seemed as if the *Luftstreitkräfte* was grounded in mourning for its greatest ace. The Red Baron himself had finally fallen to the guns of Capt

Roy Brown of No 209 Squadron on the 21st. His death was an enormous blow to the morale of the German airmen. Never again would the enemy fight with quite the same confident fire.

By the beginning of May the panic was over and the German advance was finally brought to a halt by determined resistance. Unlike the previous April, the RAF's losses had been mainly from among the fighter squadrons. During the German breakout, 130 fighters had been lost. No 1 Squadron had been in the thick of it for five weeks and suffered one killed, one missing, five wounded and five captured. This was light by comparison with the casualty lists of some units, but it still entailed a turnover of well over half the pilots in one month.

May saw a concerted Allied effort to achieve air superiority. In the first week Clayson continued his rapid climb up the scoreboard and Capt A. A. Rigby almost collided with an enemy two-seater he was attacking. Mills bought down a scout at Kemmel on the 7th and, although wounded, finally despatched it down among the treetops. A week later Clayson, now promoted to Captain, opened fire on an Albatros DV at long range before closing to 150yd for a second burst. Its starboard wings folded back and it went into a flat spin, belching smoke from the cockpit area before falling to pieces in the air. Lt Ernest Owen was another new name on the lists. He had shared a Pfalz with Mills on the 3rd, and got one of his own a few days later. In debit, the Squadron recorded the capture of Capt C. C. Clark and 2nd Lt J. C. Wood on the 8th.

The following week brought the first real lull in the fighting for some time. It was greatly welcomed at Clairmarais, with all hands grabbing what sleep they could before another burst of activity in the final days of May. Returning refreshed to the fray, No 1 Squadron then experienced one of its most successful weeks to date: for no loss it claimed seven kills and four driven down out of control. Clayson was really coming into his own by now, claiming three victories and sharing in the destruction of a similar number. Lts Kelly, Owen, Kullberg and Scroggs also featured prominently in those kills that were shared by a whole patrol.

Capt K. S. Henderson went missing on 2 June, and Lt Scroggs was wounded on the 5th in a patrol action from which 2nd Lt H. S. Hennessey failed to return. They were avenged the following day when Capt Clayson led his patrol against a large formation of Pfalz and Albatros Scouts to destroy four without loss. Clayson himself downed two of these and Lt Owen got a Pfalz. For Lt C. B. Henderson revenge was particularly sweet, although the sight of a flaming Albatros was small compensation for the loss of his elder brother earlier in the week. Three days later Clayson led 'B' and 'C' Flights against three Triplanes over Armentiers, forcing one to land and tip up on its nose on the Allied side of the lines. Its fin now adorns the Squadron's museum at Wittering.

Throughout the summer combats of 1918 the same names recur like a ticking clock: No 74 Squadron had Mannock, No 64 — Tempest, No 85 — Maj Billy Bishop; and for No 1 there was Clayson. There is no doubt that Clayson, like Fullard, would have ranked with the highest scorers had time and circumstance not been against him. As it was, he destroyed 15 German aircraft between February and July 1918, and many more were brought down by the pilots that he was leading. Of the top scorers of the Great War, only Bishop was a teamwork man. Mannock and Ball fought alone by choice, and Richthofen got most of his 80 kills against lumbering two-seaters. Both of No 1 Squadron's outstanding fighter pilots, however, were also exceptional team leaders, and contributed greatly to their cause by so being.

June was a generally dull month, both in weather and action. Clayson notched up a Pfalz and a balloon on the 15th and led 'C' Flight to a joint victory on the 28th, plus a further Albatros for himself; but otherwise, low clouds and periods of rain kept both sides firmly fixed to their waterlogged airfields for most of the month. The first list of awards of the new Distinguished Flying Cross was announced on the 23rd, and it was not surprising to find that it included one Capt B. J. Clayson.

On 2 July he was at it again, attacking a Hannoveran over Armentieres. On opening fire, the enemy machine suddenly went into a steep dive. The observer's map fell overboard, quickly followed by the observer himself, who desperately caught hold of the gun mounting and lay along the top of the fuselage as the aircraft continued its headlong plunge. His efforts were in vain, for the machine burst into flames and finally crashed. Lts C. J. Bateman, H. A. Kullberg and W. A. Smart shared a victory on the following day, which was added to by a Kullberg solo effort on the 12th.

Enemy aircraft were now becoming increasingly difficult to find and engage. The RAF communiqués for late July tell the same story

Left:
A patrol of SE5As leaving Clairmarais in June 1918. The wreckage by the hangar is of a No 1 Squadron SE5 which had collided with a machine from No 33 Squadron. *Squadron Archives*

Bottom:
Clairmarais, July 1918. The original tail-to-tail photograph. The pilots are (left to right): Lts Clayson, Newman, Smart, Nesbitt, Knight, C. B. Henderson, Crossley, Bellin, Mawby, Capt Simpson (with scarf), Capt Ira Jones (behind fin, visiting from No 74 Squadron), Capts Owen and Lavers, Lt Simpson, Padre Banks, and Lts Bateman, Forsyth and Kullberg. *Squadron Archives*

day after day: 'Very few EA seen', 'No combats took place', 'Enemy aircraft not particularly active'.

The Allied armies that had reeled before the German blows of March and April were now readying themselves to strike back. The Americans too were by now well and truly in the war; in addition to several American fighter squadrons attached to RAF headquarters, many pilots had been posted to individual British squadrons. No 1 had its fair share, and Lts R. H. Ritter, H. A. Kullberg, G. R. Touchstone and the other attached members of the United States Air Service fitted in well, with their eagerness for action and their easy-going personalities.

Capt Clayson was posted out at the same time as Maj Barton-Adams, and the command of No 1 passed to Maj W. E. Young. The new CO was an experienced and professional fighter leader who disagreed emphatically with the old policy of non-flying Squadron Commanders — if Bishop and Mannock led their units in the air, so would he. He had no intention of wasting the experience he had built up during two successful years with Nos 60 and 74 Squadrons, and he flew at the head of Squadron patrols whenever his administrative work permitted.

On 5 August, two days after Maj Young took over, the Squadron was moved up from Clairmarais to Fienvilliers in readiness for the final Allied offensive that started on 8 August. Here the Squadron was moved into IX Brigade, and came under the orders of OC 51st Wing, RAF.

The Squadron was immediately committed to battle as part of a force of 480 fighters employed in low level softening-up attacks in preparation for the Allied advance. The first day was hectic. Repeated attempts were made to destroy Peronne Bridge, some 20 miles over the lines, under a low cloud base, and two machines were lost. During these sorties, however, Lt Ernest Owen claimed an Albatros DV and Lts Kullberg and Pegg added to their growing scores. Lt K. C. Mills was killed in a crash after being hit by groundfire and one of the newly-arrived Americans, Lt Touchstone, was shot down and captured. A few days later the task was switched to that of escorting bombers to targets deep behind enemy lines. Little action was seen for the next fortnight as the enemy scouts seemed loth to attack heavily escorted formations. This did not please the new CO. He said in some notes on the work of the Squadron:

'Naturally, this work gave little opportunity for getting Huns, and the few offensive patrols that we did, probably not more than six in the final three months of the war, formed a very welcome change. The escort work, however, was carried out very successfully, only one machine being lost in 20 raids, and that went down either hit by AA fire or through engine trouble'.

One variation on the escort theme was provided from 23 to 25 August, when No 1 was attached to the 13th (Army) Wing for ground strafing duties in support of the Allied advance. Four squadrons were later commended for these operations by RAF headquarters. Coincidentally, the other three were subsequently to become No 1's closest friends and rivals in the coming decades of peace and war — Nos 43, 54 and 73 Squadrons. Lts D. Knight and C. G. Pegg brought down enemy aircraft during the three days of work, but another of the Squadron's American friends was lost when Lt Ritter failed to return on the 24th.

After this short spell of excitement, the drudgery of close escort work went on uneventfully until 16 September, when the Squadron, escorting a formation of DH9s from Nos 98 and 107 Squadrons to Valenciennes railway station, met enemy scouts in force just west of Marquion. A running fight ensued to the target and back and most of the SE5 pilots were engaged with one or other of the enemy machines. Kullberg, who had scored again the previous week, eventually shot one down in flames and, finding five others on his tail, was forced to spin down and 'contour chase' the whole way back to the lines. The enemy followed across for once and he was wounded three times before making a safe landing at Fienvilliers. Nevertheless, combats of this sort were becoming increasingly rare as much of the fight had now gone out of the German scout squadrons. The end was obviously not very far off.

For the next month the Squadron ranged into enemy territory virtually unopposed. The pilots sat for long hours on the wingtips of the heavy DH9s, laboriously forming up and climbing for height over the bombers' base airfield before setting out at 10,000ft to chunter eastwards in a sky pocked with the angry puffs of Archie. Some of their targets were well behind the lines: Aulnoye, for example, was 30 miles over, while Marienbourg, attacked on 30 October, was nearer 40. As much of their fuel was used up before even setting course, there was never very much to play with. 'It was fortunate', said Maj Young of one such raid, 'that no Huns were encountered as the petrol supply would not have allowed time to stay and fight — as it was, practically all machines scraped back on their emergency tanks.'

Although the Squadron moved up twice more during October, to Senlis on the 6th and Bouvincourt on the 26th, it was all becoming a little boring for the more experienced pilots, and highly frustrating for the new arrivals, who feared it would all be over before they too could experience air combat. Only five kills had been recorded in the previous month. In addition to Kullberg's final victory, Cameron, Pegg and Moody had scored and one of the new Flight Commanders, Capt W. Pallister, had claimed a Fokker DVII. But now it all seemed to be over. All they asked for was one final, glorious dogfight.

The day of 29 October dawned beautifully. The visibility soared into the blue morning with only a few light cumulus clouds to break up a perfect day. As the sun took off the autumnal chill, the SE5s were wheeled out of their canvas hangars and lined up on the field to be loaded with ammunition and given that final loving polish.

The operation order read:

'All available machines for offensive patrol in conjunction with bombing operation of No 107 Squadron on Marienbourg Railway Junction.

'Leave ground: Capt Lavers 1300
Capt Hoidge 1350

'Duty: Squadron will proceed to the area ANOR-TRELON-LIESSIES which they must reach by 1400 hours, and will patrol this area with a view to escorting the bombers back to the lines.

'Height in Area: 12,000ft to 14,000ft.

'Note: If the bombers are not seen after the area has been patrolled for about 10 minutes, an ordinary OP will be carried out.

'Capt Lavers' formation will follow Capt Hoidge's formation throughout.

'Signed *W. E. YOUNG* (Major)'

Capt Pallister was not flying that day so 'B' Flight was split between Lavers' and Hoidge's formations. On the dot of 13.00hrs the first flight of six aircraft lifted their wheels into the clear October skies.

They never did meet those bombers. After orbiting the RV for some time, fruitlessly seaching the skies to the east, Hoidge divided the Squadron into smaller sections for offensive patrols. One of these was led by Lt N. T. Trembarth, who subsequently reported:

'While leading an offensive patrol, I observed five Fokkers leaving their aerodrome near Hirson. I circled into the sun and attacked them at 10,000ft. During the course of the engagement, I attacked one machine and sent it down in flames. I was then attacked by four others who punctured my petrol tank and shot my elevator controls away, sending me down out of control. I regained control at about a thousand feet. About 100ft from the ground I stalled and shot at another machine which crashed, I myself crashing very close to him. I was subsequently informed by the German Squadron Commander who visited me in hospital that the second man had been shot through the head'.

Meanwhile, another section had joined in. Capt Hoidge shot down one of Trembarth's assailants while Lts Cameron and Moody claimed another DVII. On the way home the pilots were bounced by a further formation of enemy scouts, one of whom riddled Lt W. Newby's aircraft and sent him spinning to his death. Lt W. A. Smart immediately snapped in on this Fokker's tail, and one short burst was sufficient to chalk up the Squadron's final kill of the war.

Trembarth was not to remain in captivity for long. All that now remained were six uneventful escort missions, for on the morning of 11 November 1918 Maj Young received word that operations were to be cancelled until further notice. In a note to Sqn Cdr Capel of No 201 Squadron (the former Naval No 1 Squadron with which they shared the airfield), Young wrote on behalf of his colonel:

'Your patrol this morning is washed out, but the colonel wants you to continue training . . . get everything in order, stores cut down, etc, in case we move at short notice. I believe our destination is COLOGNE!!! Seems too amazing to be true, doesn't it?'

But Maj Young was never destined to lead his men on to Cologne. A week after the Armistice they moved to Le Hameau, where, itching to go home, they spent a frustrating couple of months. During January and February of 1919, they were given the job of ferrying SE5s to England; nearly 100 machines were flown home to end their days in less glorious funeral pyres than those of their many victims.

Towards the end of February Maj Young was posted home, together with many of the pilots and groundcrew. The remaining cadre, commanded by Capt Stewart Lavers, followed on 1 March. The shadow of the former pride of the line arrived at London Colney where it twiddled its collective thumbs until October, and was then moved to Uxbridge — 'for disposal'.

For a while, the future of No 1 Squadron hung in the balance. So many proud squadrons were no more, and the fear of total disbandment was ever in the minds of the remnants. The RAF itself was only still in being because of the tenacity of Trenchard, but it was unthinkable that there should be an air force whose squadrons did not include a No 1.

Happily, a need was now seen for air power to protect the trade of a still powerful peacetime Empire, and it was decided to give the numberplate of the Squadron to a new unit about to form in India. Capt Lavers handed over his unit's final papers on 20 January 1920, without fuss or ceremony, and went his way. 'Number One is dead — long live Number One'.

CHAPTER FOUR

Foreign Parts

Shortly before the Christmas of 1919, as the pale shadow of No 1 Squadron still kicked its heels at Uxbridge, Sqn Ldr J. B. Graham MC was summoned to the Air Ministry. There he was delighted to learn of an immediate posting to India to form the new Squadron that was to rise like a phoenix from the ashes of the old. In the struggling postwar RAF, flying appointments were rare for men of his rank. He had flown in the Australian Flying Corps during the war, and had stayed on in England to make the RAF his career. Given the opportunity of selecting a proportion of his officers, Jack Graham was overjoyed, and before catching the boat to Bombay, he dashed around making contact with old comrades. At the RAF Club he bumped into Fg Off Geoffrey Pidcock, who had been attached to No 1 in 1917, and who readily agreed to join the Squadron when available later that year. For a Flight Commander he chose the likeable and headstrong Flt Lt Frank Soden.

While Sqn Ldr Graham was travelling to India, a cadre of the Squadron was assembling at Risalpur under the senior Flight Commander, Flt Lt Geoffrey Robarts, and the administrative officer, Fg Off Charles Knight. The arrival of the CO was followed swiftly by that of the first of the Squadron's Sopwith Snipes. Powered by a Bentley rotary engine of 230hp, the Snipe could cruise at 120mph at 10,000ft and was armed with twin synchronised Vickers guns. It handled like a dream, and was to be flown by No 1 Squadron for the next six years.

No sooner had the first aircraft been assembled from its crate than the Squadron was ordered to move down to Bangalore in Madras, where it was to share the airfield at Gaza camp with No 3 Squadron. The Squadron arrived at Bangalore on 17 May 1920 and set to work immediately to assemble the aircraft and settle into its own Mess. The aerodrome was commanded by Wg Cdr George Pretyman, who had led No 1 in 1915. Now ably assisted by his delightful Irish wife, he made sure that his old unit was made welcome in its new surroundings.

Very little flying was accomplished in the early days at Bangalore. The locals were law-abiding, and despite the Moplar rebellion in Ceylon, the squadrons were not called upon for any operational tasks. There was also the problem of spares. Because of the immense distance from the Air Ministry and the financial stringency of the Government, replacement parts were virtually non-existent. As a result the Snipes spent most of their time firmly on the ground. At one time the situation was so bad that the stores officer, Fg Off William Bowden, actually had to go down to the bazaar to buy his split pins, nuts and bolts. To stave off the boredom the pilots went to great pains to cultivate good relations with the army units at Bangalore, even to the extent of completing a drill course with the 2nd Dorset Regiment. Their particular friends were the Scots Greys, who put a number of the pilots through their riding school. Fg Off Pidcock, who joined the Squadron in June, took this opportunity. He recalled that he must have become a good horseman eventually, as he fell off in every possible direction.

On 18 June the Squadron was visited by the Air Officer Commanding India, Air Cdre T. I. Webb-Bowen. This was not a happy event. At about the time that No 1 had arrived in India, the AOC had issued an order that cut the pay of all junior RAF officers in India very considerably. The reason for this was that he considered that they had too much money and were spending too much on drink in the Mess. In the very same order, his own pay had been increased, and the Air Commodore was regarded with great disrespect as a result. To add insult to injury, the order was cancelled a few days after No 1 Squadron eventually left India.

The life of a young RAF officer in India was often fairly miserable. The cost of decent living was very high and, with little flying available, even the close comradeship of a fighter squadron was little solace for the lack of young mixed society. It is small wonder that junior officers in India often drank too hard for the strength of their pockets, and some were cashiered for bankruptcy as a result. The main gathering places in Bangalore were the Gymkhana Club and British United Services Club. These were frequented mainly by the army and members of the civil service, who were generally far better off than the RAF.

The most expensive item in an officer's kit in those days was an immaculate pair of field boots, whose leather was polished to a stunning brilliance for the frequent parades. The already unpopular AOC now ordered pilots to wear these boots for flying, in total ignorance of the fact that the Snipe's engine had an unfortunate habit of spewing castor oil liberally around the cockpit.

In the face of these tribulations the Squadron's morale was surprisingly high, which reflected well on the quality of its leaders. Flt Lt Robarts was posted out and replaced as Flight Commander by Flt Lt Francis Vincent. Vincent was as tough as they come, having served in the Royal Marines during the war, but his outstanding sense of humour made him universally popular. He used to keep a lion cub in the Mess at Bangalore, which was everybody's pet until it got so big and unruly that he had to send it back to the zoo at Mysore.

In March 1921, the Colonial Secretary, Mr Winston Churchill, presided over a conference in Cairo that decided upon a policy of 'air policing' for the protection of the British mandated territories in the Middle East. It was decided that Mesopotamia could be policed by five squadrons of DH9s and Bristol Fighters, reinforced by two squadrons of troop-carriers and one of single-seat escort fighters. This would enable the ground forces to be reduced from 13 battalions to one brigade, supported by several companies of RAF armoured cars.

No 1 Squadron was chosen to join this force, and the news was received at Bangalore with predictable delight. In the two short weeks available the aircraft were crated and made ready to move. Fg Off Bowden organised the packing of his stores, aided by the excellent FS Preen. On 17 and 18 April the Squadron members gave a farewell concert party — calling themselves the 'Astrals' — which was a resounding success, and on the 20th they all boarded the train for Bombay.

The voyage to Baghdad came as a welcome holiday at sea. They left Bombay on 27 April 1921, aboard the SS *Brandenburg*, and after calling at Karachi they arrived at Basra, where they transferred to a stern-wheel tug for the slow paddle up the Shatt-Al-Arab. The aircraft and stores were loaded on two barges strapped to the side of the old stern-wheeler, which served as excellent buffers whenever the coxwain ran aground on the sandbanks.

On arrival at Baghdad, Fg Off Pidcock, who was by now the Squadron's adjutant, set off with a sergeant and four men to start setting up the camp at the new airfield at Hinaidi, on the banks of the Tigris, and became the first RAF officer to sleep there.

Those first months at Hinaidi were hectic and unpleasant. With the aerodrome and permanent accommodation still under construction the Squadron spent the first hot season under canvas. Every drop of water had to be brought in by bowser, and, with no ice or punkahs available, the blistering heat of midday became almost unbearable. Under

Two Sopwith Snipes taking off from Hinaidi.
Sir Dermot Boyle

these conditions sickness was inevitable, and therefore Flt Lt F. G. Murphy of the Medical Branch was posted to the Squadron to cope with the many cases of sandfly fever that occurred before autumn brought some relief.

The completion of the Officers' Mess and the airmen's accommodation coincided with the cooler weather to make life more enjoyable. Local servants were hired, and the Squadron's catering was put out to contract to a succession of local rogues, who were replaced regularly as new fiddles came to light. Improvements to the aerodrome made possible some intensive flying, to the relief of the underemployed pilots.

By now the Squadron had been welcomed by the European community of Baghdad. A syndicate of pilots bought a motor boat and several of them took up polo under the guidance of 'Spud' Murphy. Now and again they would hire a train and travel 40 miles up-river to Lakuba, where there was some excellent duck shooting to be had. Without the financial embarrassments suffered in India, they were really quite well off. Within a year most of the officers had paid off the overdrafts that they had run up at Bangalore, and were able to buy a car and a gun, a couple of polo ponies or a share in the Squadron's motor launch.

Hinaidi rapidly became the hub of air power in the Middle East. As the complement of its squadrons built up to a maximum of eight, each with its own messes, hangarage, identities and social trademarks, the base became a hive of bustling activity, both in the air and on the ground. The other fighter squadron was No 6, equipped with old two-seat Brisfits that were scarcely a match for the manoeuvrable Snipe. No 45 Squadron, together with No 70, flew Vickers Vernon troop carriers; their task was to deploy the army to any incident which could not be contained by close air support from the fighters and four squadrons of DH9A light bombers.

The years of 1922 and 1923 were, for No 1, composed largely of pleasant, unhurried tropical training routine. The lack of suitable airfields up-country, the short range of the Snipe and the unopposed nature of the operations being conducted by the light bombers, all left little work to be done. This is not to say that they were idle. Constant practice in the art of ground strafing at targets set in the desert, in combat, and in the dropping and picking up of messages honed their flying skills to a fine degree.

As the months passed, the faces came and went. In November 1922, Sqn Ldr Graham was replaced by Sqn Ldr G. G. A. Williams, who commanded No 1 until October 1923. He in turn handed over to Sqn Ldr E. O. Grenfell, himself no stranger to the unit. He had flown on its very first offensive operation in France in March 1915, and said that he was very glad to be back. Willy Fry, who had been on the Squadron for some time, was promoted to take over one of the Flights, while Francis Vincent was posted out and was replaced by Flt Lt Francis Luxmoore as senior Flight Commander. Geoffrey Pidcock was also posted, but he volunteered to stay on at Hinaidi and took over the MT pool. During his tour he had been known as Piddling Officer Flycock, but his career prospered and he later became Piddling Leader Squadcock, eventually retiring as Air Pid-Marshal Vicecock!

Two Snipes of No 1 Squadron with a Vickers Vernon troop-carrier from No 45 Squadron at Hinaidi.
Squadron Archives

Sqn Ldr E. O. Grenfell MC, DFC, AFC, OC No 1 Squadron in Iraq from 8 October 1923 to 25 May 1924, and at Tangmere from 19 March 1928 to 27 July 1931. Eustace Grenfell had taken part in No 1's first offensive operation from St Omer on 12 March 1915. *Squadron Archives*

The Army Commander had been succeeded by AVM Sir John Salmond on 1 October 1922, who was gazetted as a temporary Major-General to give him full powers over the remaining army units. Sir John was the brother of Geoffrey Salmond, No 1 Squadron's first operational Commander, and had himself led 2nd Wing RFC for a time when No 1 had been attached to it at Bailleul. The AOC's presence was not the only echo from the past at Hinaidi, for Guy de Dombasle was also serving on the station as a Wing Commander.

In addition to its full complement of Sopwith Snipes, the Squadron was issued with three Nieuport Nighthawks for experimental purposes. The Nighthawk was a rare bird indeed, for only four examples of the type ever saw squadron service with the RAF. Two of them, HR8544 and JR6925, were powered by 325hp Armstrong-Siddeley Jaguar engines, while JR6926 had a 385hp Bristol Jupiter. The Nighthawk was historically significant because it was the first RAF fighter to be powered by a stationary radial engine: all its predecessors had been driven by rotary powerplants. The presence of an 'R' in the serial numbers of the Nighthawks indicates that they had been rebuilt, by Gloster, after the Nieuport company had gone bankrupt in 1920. They were pleasant to fly, but the engines were a complete disaster. They suffered badly from sand ingestion, and caused the type to be withdrawn from service by the end of 1923.

Throughout 1923, life was fairly quiet because the Turkish irregulars who had been stirring up the Kurdish tribes in the north of the country now withdrew, and the border dispute quietly died away. A new source of trouble was not long in emerging. Sheikh Mahmud, the Governor of the Sulaimaniya district, had been exploiting Kurdish grievances since 1919, and he now began to incite open rebellion against the British mandate. The Kurds, a fierce, proud and warlike people of the northern mountains, had never been occupied, but they objected to the dominance of the Arab lowlanders in the Iraq Government, and to paying taxes towards their upkeep. The British administration, under a League of Nations Mandate to keep the peace, was bound to discourage their dissent, and to supress it with force where necessary. The principle of 'air policing' was an effective, economical and humane way of doing this. If a tribe or village was misbehaving, by raiding other villages and carrying off their women or by refusing to pay taxes, they would be banished from their own village for a while. To deter their return, a Flight of Snipes would occasionally be sent to spray the empty village with gunfire. This seemed to work, because the pilots never saw any villagers or livestock on these sorties.

Back at Hinaidi, Sir John Salmond handed over Iraq Command in April to AVM Sir John Higgins, and a few weeks later Sqn Ldr E. D. Atkinson DFC, AFC, arrived to take over command of No 1 Squadron from Eustace Grenfell. Edward Atkinson, known to one and all as 'Spider', did not have a happy tour with the Squadron. He was in poor health when he arrived and his condition was not helped by the climate or his partiality to whisky. As a result, a great deal of extra responsibility was thrust upon his Flight Commanders. Flt Lt Joseph Fall was second-in-command at the time and bore this burden with great credit, being well supported by those, like Fg Offs Donald Anderson, Edwin Mattock and William Shaylor, who had served with No 1 since its arrival in Iraq.

Another source of friction with the Kurds was the use of levies acting as ground forces in

Flt Lt Oswald Gayford leading the Squadron over the Tigris near Hinaidi. *A. E. Garnett*

place of British troops. On 4 May 1924 a demonstration was broken up by the levies in a Kurdish bazaar in Sulaimaniya. Severely outnumbered in the riot that followed, the levies opened fire on the crowd, inflicting heavy casualties before they withdrew. This was just what Sheikh Mahmud was waiting for. Seizing upon the furious reaction of the Kurds, he declared a holy war — a Jihad — upon the British and the Baghdad Government. His first campaign was fairly successful and he succeeded in denying much of the northern area of Iraq to the British forces. This state of affairs lasted until July, when a force of 42 light bombers struck at the town of Sulaimaniya and pacified it without loss of life — a remarkable demonstration of the success of air policing. The Sheikh then withdrew into the hills, allowing the squadrons to operate once more from Sulaimaniya airfield.

With the campaigning season over, Hinaidi returned to normal. No 1 had not been involved in this first series of skirmishes, as the Snipe lacked the necessary range to engage targets in the hills without a forward base.

Joseph Fall was posted out in December, and his place was taken by Flt Lt Oswald Gayford, one of the best known characters of the peacetime RAF. Gayford was immensely popular and proved to be a first class administrator. He sharpened up the Squadron organisation very considerably during his tour and was very keen on improving relations with the other squadrons and the commercial community in Baghdad. As a result, the No 1 Squadron Officers' mess became quite a social centre. The Mess tennis court was always well patronised and impeccably maintained — even the ball boys were provided with sparkling white kit from Mess funds. There was never a shortage of local urchins to fill these posts, as they were given free lemonade and a salary of 2d per day. The Mess was well run by Sgt Killick, and the officers' personal bearers were always in attendance to serve their masters at meals. Guest nights were held twice every month; a large beer cost 11d and a John Collins was sixpence.

Their daily routine was a pleasant round. The Squadron would be up bright and early, and most of the flying was done in the early morning when the air was cool and still. Aerobatics and mock attacks were practiced daily. There were few restrictions on low flying and one pilot dipped his wheels into the Tigris on one occasion, fortunately recovering the situation before he nosed over. The aircraft would be back in their hangar by lunchtime, after which there was time for a siesta, a swim or a quiet game of tennis before changing for dinner or an excursion to Baghdad where the Squadron were members of the Al-Wiyah Club. Each squadron had its own Mess and held open house on different evenings of the week to provide a continuous round of parties.

Sooner or later, a chap could tire of this exhausting existence, and then it was time to take a spot of leave in Cairo. This was achieved by hopping on a mail flight in a DH9A. The trip would take them down the Imperial Airways route over Transjordan and Palesine, following a furrow ploughed in the desert as a navigation aid.

In the spring, Sheikh Mahmud once more began to stir up his tribes, and a series of incidents in the north called for action by the RAF. This time the Squadron was overjoyed to hear that it was to send a detachment up to Sulaimaniya, commanded initially by Flt Lt Francis Luxmoore — the CO was away in hospital at the time and Oswald Gayford was required to stay at Hinaidi, much to his disgust, to act in his place. This first party to go north comprised Luxmoore and Fg Offs Adams, Young and Moon. They loaded their Snipes with an assortment of baggage including some borrowed Mess crockery, strapped their bedrolls to the outside of their fuselages and flew off sporting big grins.

Flt Lt Francis Luxmoore, the senior Flight Commander from 1923 to 1926, with his dog Raggis. The dog went everywhere with Luxmoore, who cut a hatch in the Snipe's locker so that Raggis could see where he was going. 'He travelled thus for many miles, and accompanied me on a number of operational sorties, when he would do his best to add colour to the scene by barking furiously if I had occasion to fire my guns. . . . He was also the first dog ever permanently to wear a "beer spanner" on his collar. He carried this with dignity, so that no river picnic, shooting party or the like would be marred by the lack of that essential tool.' *Country Life*

For the first few weeks of the detachment in April 1925, very little happened. Parties of tribesmen would sally forth from the hills to cause trouble in peaceful villages. By the time the Snipes were scrambled the Kurds would be safely home in their mountain redoubt. Strafing runs were made on a number of these sorties, but it was always difficult to determine if there had been any positive results. During May, however, a large force of well-armed rebels met the army in a full-scale action around Halebja. Heavy fighting developed in the area for several days and the army's efforts to take the town were not proving very successful. At a critical moment in the fighting, Luxmoore and his Flight arrived overhead and caught a large body of the enemy in the open. During the strafing attacks that ensued they routed the rebel forces in disarray. Halebja fell on the following day as a direct result of No 1 Squadron's most significant action during its stay in Iraq.

The Christmas trooping season was notable for the arrival of the Squadron's first peacetime non-commissioned pilots — three sergeants who came in January 1926. It was now policy to send only second-tour pilots to No 1 Squadron. One of these was Fg Off Dermot Boyle, the first graduate of the Royal Air Force College at Cranwell to join the Squadron.

The magnificent career which followed his tour of No 1 was to take him to the very pinnacle of the RAF when he held the post of Chief of the Air Staff with the rank of Marshal of the Royal Air Force. Sir Dermot vividly recalled his arrival on the Squadron:

'On disembarking with two colleagues at Basra, I took the train to Baghdad, where I was met at five in the morning by my Flight Commander and the Adjutant, which we thought was really red carpet treatment. We then drove to Hinaidi, and on arrival in the Mess, I was presented with a glass of beer for breakfast, which I found very depressing. However, we soon got used to this and after three or four months the beer for breakfast had no ill effects on us whatsoever'.

Sqn Ldr Atkinson was now posted back to England and Oswald Gayford assumed acting command of the Squadron — a job he had been holding down, in effect, for some time. This state of affairs was rectified in April 1926 with the arrival of Sqn Ldr Cyril Lowe MC, DFC. 'CN' had fought with No 24 Squadron in France and had rejoined the RAF in peacetime. He was also a very famous rugby winger and had

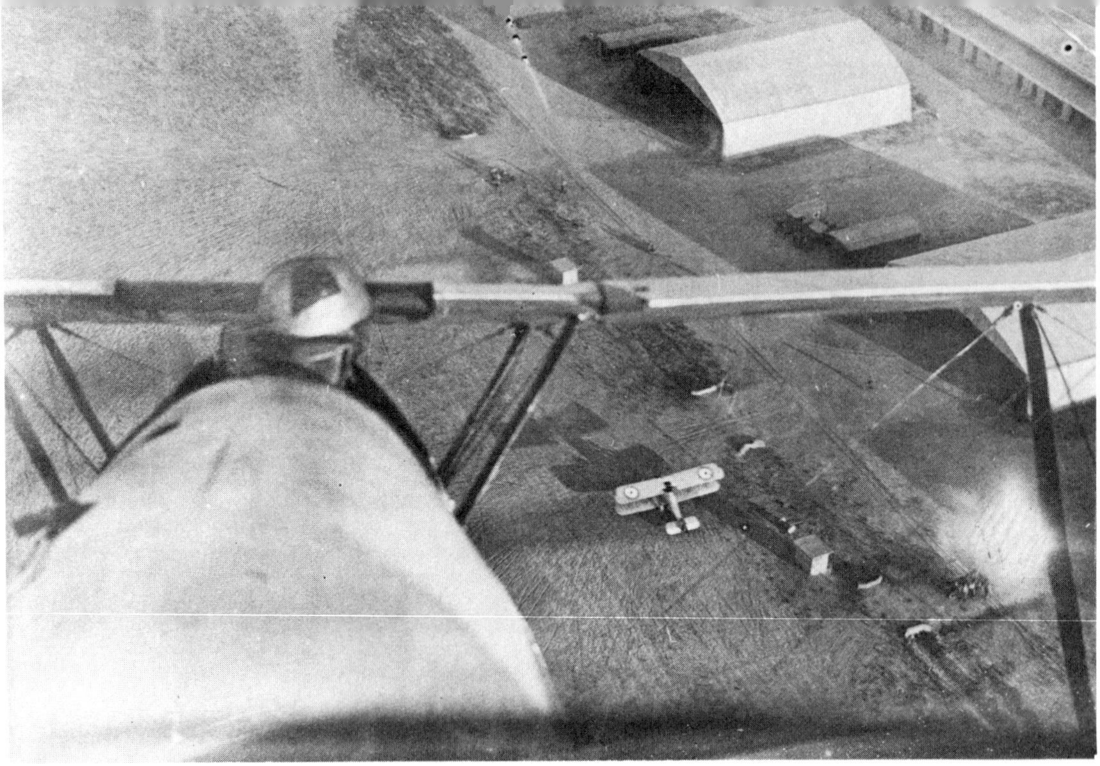

The Squadron hangars at Hinaidi, taken by a camera fastened to the fin of Fg Off Boyle's Snipe.
Sir Dermot Boyle

gained 25 English caps, a record which stood for many years. He had been the outstanding player of the RAF vs Army match in 1921, scoring four of the five tries, one of which had started from his own goal line.

Sqn Ldr Lowe's enthusiasm for flying was boundless and he encouraged the pilots to spend a lot more time in the air than before. He had recently been a flying instructor at Cranwell, where he had come into conflict with the authorities by insisting that his cadets spend the maximum time in the air, which was deemed to be detrimental to their classroom studies. His leadership in the air and unorthodox flair for command on the ground soon endowed the Squadron with a bounce and spirit that recently had been lacking.

On one occasion he decided to combine his love of shooting with a practical exercise by taking the Squadron to a landing strip some 50 miles up the Tigris, shooting duck all day, and flying home for night landings at one minute intervals. A thoroughly enjoyable day was spoilt when Fg Off Stuart Young crashed on landing and was burned to death in spite of the efforts of his fellow pilots to pull him from the blazing Snipe.

Fg Off Dermot Boyle. *Sir Dermot Boyle*

In 1926 No 1 Squadron stepped up its air-to-ground training and targets were erected in the desert 22 miles from Hinaidi — 'which shows the opinion that the locals had of our accuracy', said Sir Dermot Boyle.

This entailed a most awful journey out to the range in a closed truck, and the ground crew would go out together with 3 or 4 pilots. The journey was absolutely horrible — bumpy, dusty and dirty. Now a couple of lucky pilots, and on this occasion I was one of them, used to fly two aeroplanes out to the range. We would do our air firing, and then land in the desert. The other pilots would then have a go in our aircraft and when the thing was over we would fly our aircraft back.

'On one of these occasions, a great friend of mine came up to me during the morning and said: 'I cannot stand this lorry journey, would you give me a lift back on your aeroplane?'

'Quite light-heartedly I agreed, and went on with the business of the morning. When the time came to leave we stupidly decided that he should be accommodated, if that is the right expression, lying along the fuselage. He and I used to ride horses together, and we thought that you could ride the Snipe and hold on to the back of the coaming behind the cockpit and everything would be all right. We started up, he got in the "go" position. I didn't put my overalls on, I stuffed them behind my back, and off we went!

'The line of take-off that I chose across the desert was not very good. There was an undulation. As we came out of the undulation, there was a shriek of abuse from my passenger, whose mouth was much too close to my ear for comfort. I looked up in amazement and horror, for — if I may be Irish for a moment — he was doing a sort of handstand on his elbows, up absolutely vertically above my head, with his feet way up in the air!

'Full of confidence I shouted at him: "Don't worry. Hang on — it'll be all right when we get in the air!" Little did I know. Any aviator will know that if you take a little short chubby aeroplane, and put 14½ stone 4ft behind the centre of lift, you would be apt to cause some difficulty. And if in addition, he did a handstand in front of the rudder, you would have further difficulties. Well, the aircraft was practically uncontrollable, and I remember the horror with which I discovered this as I got into the air. I had to jam the stick against my leg to prevent over-controlling, and I gradually edged the aeroplane towards Hinaidi. I found that if I throttled back the engine the situation improved, but of course there is a limit to how much you can throttle back without incurring other technical difficulties!

Fg Off Studdert — Fg Off Dermot Boyle's unfortunate passenger. *Sir Dermot Boyle*

'We continued on our way. My poor passenger had another discomfort to suffer. The overalls, which I had stuffed rather light-heartedly behind my back, now began to emerge into the slipstream, led by the belt buckles, which started bashing my unfortunate passenger in the face, and he hadn't a spare hand with which to defend himself. We finally got back to Hinaidi and I landed at the far side of the airfield, and he got down with a broad smile. While he was walking back around this huge airfield, I went up and did some aerobatics in order to take people's attention off him.'

By the end of May the Squadron had been ordered to set up the detachment at Sulaimaniya once more. This time it was commanded by Oswald Gayford, and soon after arrival at the forward base the Flight was employed in a number of strafing missions, mostly against elusive parties of tribesmen operating from the foothills. From 17-19 June the pilots operated against rebel villages in conjunction with No 30 Squadron from Kirkuk. While diving on one of these villages, Fg Off Boyle noticed that his engine was making a peculiar noise.

'When I dived the next time, the whole cowling came off and was drawn into the engine like a mincing machine, bringing the propeller to a grinding halt. I did an ozzletwizzle and landed without breaking anything else. I then had to burn my nice E6655 by firing my Very pistol into the petrol so that it would not fall into the hands of the rebels.'

He was rescued by the levies, and spent three days getting back to base — a journey that had taken him 40 minutes on the way out!

Sqn Ldr Lowe visited the detachment soon afterwards, and had a narrow escape himself when he dropped in on a unit of the levies. He had landed alongside them in a wadi, and a levy was detailed to sit in the cockpit to keep the engine running. There were only a few hundred yards available for take off, and the CO, business completed, was a trifle concerned to find his engine oiling up as he took off. He made it over the cliff at the end with inches to spare.

Units of the Iraq levies used to conduct long patrols of the frontier, and the Snipe pilots used to make a daily check on their progress and safety. On one occasion they decided to drop some dead chickens on them, as a religious feast was due. 'Then we remembered that their religion required that they kill the chickens themselves, so we decided upon a live drop', recalled Sir Dermot. He set off with three miserable hens tied to the ledge on his dashboard with a slip-knot. The levies were

The Sulaimaniya detachment. The Snipes are parked in a camel compound on the edge of town, to protect them from Kurdish snipers. The photograph was taken from the roof of the nearby Officers' Mess, where the pilots would sleep out on hot nights. *Sir Dermot Boyle*

encamped as usual in the bottom of a deep wadi, which called for some inspired flying as he took his hands off the stick and throttle to release the first poor chicken.

'I tried to make it take off into wind, to give it the easiest possible flight. But it turned its head away and I decided to let it go. I shall never forget the sight, as the hen slowed down to 50mph and started flying. Its wings went up and up above its head, and it plummeted into the landing zone as required. I did feel an awful cad dropping the other two!'

The rebel activity began to die down as the hot season approached, but the Squadron maintained its detachment at Sulaimaniya until September.

Throughout the summer, rumours had been rife that No 1 Squadron was to return to England, and these were officially confirmed on the return of the 'Sul' detachment — the Snipes were getting old and did not have the range and weapon load to carry the offensive into the mountain homeland of the Kurds. Word was eventually received that No 1 would disband for a while on 1 November, prior to re-forming at Tangmere the following spring. A couple of the pilots were to remain on the strength of the unit to maintain continuity, but the more recent arrivals were to be dispersed throughout Iraq command and convert to other types of aircraft.

Throughout October the remaining Mess funds were used up in a series of farewell parties. Many of the officers, including Sqn Ldr Lowe and Fg Off Boyle, were posted to No 6 Squadron to fly the Bristol Fighter and went on to complete successful and enjoyable tours in Iraq. Oswald Gayford remained at Hinaidi until March 1927, supervising the wrapping up of the Squadron, assisted by Fg Off George Bucknall until the latter's departure for Tangmere. The silver and documents were carried to England in November by Fg Off Arthur Rogenhagen, and duly arrived with him at the new station in Sussex. The Snipes that had given such valiant service were ignominiously scrapped on the spot.

A pleasant and sunny era had ended for the Squadron: a greater and more glorious one was about to begin.

Fg Off Dermot Boyle aerobatting his trusty E6655 near Hinaidi. *Sir Dermot Boyle*

CHAPTER FIVE

The Golden Age

Flt Lt Francis Luxmoore had left No 1 Squadron in Iraq in 1926. After a leisurely sea trip home, a long spell of leave and his wedding, he was ordered to report to Tangmere, near Chichester, to organise the return of the nucleus of his old Squadron from Hinaidi. Soon after his arrival in February he was joined by Fg Off Rogenhagen with the silver and the documents; and later by George Bucknall, who had been helping Oswald Gayford to tidy up in Iraq. These three were all that remained of the old No 1 Squadron, and they were joined at Tangmere by seven new pilots. In April, Sqn Ldr E. D. Atkinson, now somewhat recovered from his earlier illness in the Middle East, arrived to take over from Luxmoore as Commanding Officer. By then the Squadron was ready to receive the first of its new Siskins.

The Armstrong-Whitworth Siskin was fairly representative of the early postwar biplane fighters, with a performance only slightly better than that of the Snipe. Its 420hp Armstrong-Siddeley Jaguar engine gave it a top speed of 156mph at sea level and a service ceiling of 27,000ft. It was equipped with radio and armed with twin synchronised Vickers guns firing through the propeller arc.

Tangmere, nestling between the South Downs and the sea, was a delightful spot. Surrounded by green fields and pleasant villages, close to golden beaches and a hundred country pubs, it was the perfect place for the Squadron to shake the dust of the desert from its shoes. Its role in the Fighting Area of the Air Defence of Great Britain (ADGB) was to protect the port of Southampton and the nearby naval base at Portsmouth. The other unit at Tangmere was No 43 Squadron, equipped with Gloster Gamecocks. When No 1 became operational on the Siskin, there began a period of intense friendly rivalry that was to last for over 12 years. In that time the two squadrons were to transform their small airfield into the best known fighter station in the country. In a lyrical period when the RAF was often described as the finest flying club on earth, Tangmere was to reign supreme. The golden age of the RAF undoubtedly dawned there in the spring of 1927.

Even the pilots serving on No 1 Squadron at the time would admit that No 43 Squadron was a good outfit — the second finest in the business. It had a first class CO in Sqn Ldr Brooke, and a tremendous squadron spirit. The pilots called themselves the 'Fighting Cocks', after their Gamecock aircraft, but this nickname was soon changed to the 'Chicken Farmers' by the men of No 1. These were the days when the crack fighter squadrons of the RAF painted the wings and fuselages of their silver aircraft with bright and colourful markings. No 43 had already adopted a bold black and white chequerboard pattern of which it was very proud. In reply, No 1 adorned its Siskins with two scarlet bands that traversed the full span of the upper mainplanes, and which were repeated down the fuselage sides, meeting just forward of the tail unit.

No 43 traded in its Gamecocks for Siskins in December, and had a stroke of very good fortune in the choice of their new CO, for Sqn Ldr Brooke was replaced by none other than Sqn Ldr C. N. Lowe, who had just returned from Iraq.

By the end of 1927 Spider Atkinson's health had deteriorated once more, and Flt Lt Stanley

Sqn Ldr Eustace Grenfell taking off from Tangmere in his Siskin. *Squadron Archives*

Vincent took over as acting CO in January 1928. Sqn Ldr Atkinson was later replaced by another former squadron commander, Sqn Ldr Eustace Grenfell. Although a fine pilot in his younger days, Grenfell did not now fly very often, and the leadership of the Squadron in the air was conducted by Stanley Vincent.

As summer approached, Sqn Ldr Grenfell entered No 1 for the annual Air Pageant at Hendon. During the display (the ninth in the series), the Squadron was scrambled in company with No 111 (Siskins) and No 32 (Gamecocks) to intercept an incoming wave of DH9s from No 605 (County of Warwick) Squadron. Fg Off Hugh Loch was entered in the altitude race and took third place with a height of 14,000ft. In a very minor way No 1 Squadron had reappeared in the public eye.

In the depressed days of the late 1920s and early 1930s there were few young men as fortunate as the pilots of Tangmere, which was the only station in the country to work to a tropical routine during the summer months. This involved an early start to the day, and the first sections would take off soon after 07.30. Intensive flying would than take place all morning, until the station stood down at 13.00hrs. The system had its critics, mostly of a jealous nature, but they were silenced by the fact that Nos 1 and 43 regularly achieved more flying hours than any other RAF fighter squadron — and still took the afternoon off.

After lunch the Squadron would go its several ways. Sqn Ldr Grenfell would head for Itchenor, where he kept his new International 14ft dinghy. He taught several of his pilots to sail and was known to tip the CO of No 43 into Chichester Harbour on several occasions. For the others, there was good bathing at West Wittering, village cricket, a game of golf at Goodwood or a visit to one of the many civilian friends that the Squadron had made since its return from Iraq. After dinner there was always the likelihood of a trip down to the 'Old Ship' at Bosham, or one of the many other delightful pubs in the area. Not far down the coast lay Brighton and those not otherwise engaged would descend on the unfortunate resort for a Saturday night revel that usually included a few talent-spotting drinks at the Metropole, and a visit to the Hippodrome. Beer was cheap and plentiful, and few of the locals were hostile.

In the spring of 1929 No 25 Squadron joined Nos 1 and 43 in forming No 3 (Fighter) Wing. Also equipped with Siskins, No 25 was based at Kenley, from where it joined the Tangmere squadrons for the purpose of exercises and displays. At the Hendon Air Pageant that summer they performed together as a Wing for the first time, and drew the following comment from *Flight*:

'The 27 Siskins, after taking off in formation — a truly magnificent sight — flew back and forth across the aerodrome, two squadrons from opposite directions and one from right angles, all three meeting and crossing one above the other over the centre of the aerodrome. This was accomplished with remarkable precision'.

In August each year the Squadron would prepare to fly away for its annual air firing

The Hawker Fury was the first fighter to exceed 200mph in level flight, powered by a 525hp Rolls-Royce Kestrel. This is Sqn Ldr Charles Spackman's aircraft — his CO's flag can be seen on the fin. *Squadron Archives*

practice camp, normally held at Sutton Bridge near the Wash. Although it involved a couple of weeks of very hard work, the annual camp was conducted with a pleasant holiday atmosphere. The pilots would vie with one another for the highest scores in gunnery on a target towed by a Bristol Fighter, and in steep angle strafing attacks on targets set up on the coast. Cups and trophies were awarded for the highest averages; the only real problem was the appalling serviceability of the Vickers guns. The annual camp soon became a highlight of the year's training programme. Its relaxed atmosphere was excellent for morale. It enabled officers and men to get to know each other much better, and provided a welcome rest from the frequent station duties and 'bull' that were very much a part of Tangmere's life.

In July, Eustace Grenfell was promoted to Wing Commander, and relinquished his command of the Squadron to Sqn Ldr Charles Spackman. Unlike his predecessor, the new CO spent a lot of time in the air with his men. The establishment of the unit had now grown to include some 20 pilots, of whom four or five were sergeants, and this steady growth had necessitated splitting the Squadron into two flights by 1929. 'B' Flight had been commanded initially by Flt Lt Eric Broadberry MC, who handed over to George Simmons in April 1930. Stanley Vincent, the senior Flight Commander, was promoted in the same list as Sqn Ldr Grenfell, and handed over 'A' Flight to Flt Lt Oliver Carter. For a while a third flight was formed, commanded by Flt Lt Ian Bertram.

Tangmere was commanded until November 1931 by Wg Cdr J. B. Graham, who had re-formed No 1 Squadron in India in 1920. Below him at Station Headquarters were eight officers who ran the entire station administration with a small number of clerks and storemen. Air Traffic Control did not exist and flying discipline in the circuit was dictated by individual airmanship rather than by rigid procedure. The only concession to law and order in this respect was the provision of a duty pilot, armed with a Very pistol, who lived in a hut discreetly situated between the two squadron hangars.

In February 1932 two pilots set off for the Hawker factory at Brooklands to collect No 1 Squadron's first Furies. Conversion to type was achieved by a quick chat from George Bulman, the Chief Test Pilot, and by the delivery flight to Tangmere. Under the polished alloy skin of Fury beat the heart of the true Hawker thoroughbred. The controls were beautifully harmonised and the all-round vision was excellent. The tight little cockpit was just as cold as the Siskin's, but the Fury soon found an unshakeable place in the hearts

of its pilots — it was, after all, the fastest aircraft in service anywhere in the world. For some years the Fury was to be the only fighter that could catch its stablemate, the Hawker Hart light bomber.

Within a few weeks the pilots were stunting their Furies over Sussex with seeming ease. As the spring wore on to early summer, they started working up for the Hendon Air Pageant. No 43 was going to be performing its 'roped show', so Sqn Ldr Spackman decided to enter a pair of aircraft to perform synchronised aerobatics, backed up by a formation team of three. After a month or so of competition and practice, Fg Off John Davys was picked to synchronise with Sgt Plt Colin Scragg. The aerobatic team, led by Flt Lt Oliver Carter, was also to comprise Fg Offs Thomas Hope and Barney Keily. At last all was ready, and 25 June saw five immaculate Furies lined up at Hendon.

'The weather was a curious mixture. There was a good deal of sun, mixed with a fair amount of cloud. A fresh and rather bumpy breeze was blowing, but at the same time the air was a trifle oppressive and stuffy. No rain fell, and on the whole there was little cause for complaint. Royalty was present in the person of Air Vice-Marshal HRH The Duke of York. An enormous crowd testified to the continued popularity of the display. The flying and organisation were magnificent. There was nothing to detract from the pure enjoyment of a glorious holiday in the open air.'

Above:
Sqn Ldr Charles Spackman leading eight of the Squadron's Furies over Sussex in 1933. *Flight*

Below:
A 'Flying One' formation of Furies near Tangmere. This was taken by John Yoxall of *Flight*, flying in Hawker's own Hart, G-ABMR. *Flight*

How well *Flight* captures the mood and excitement of the day! No 43 were there of course, rivals at Tangmere and feted veterans of Hendon. A host of other units were present. No 9 Squadron bombed targets with Virginias and No 16 disgorged troops from its new Atlases; No 23 brought its Bulldogs. There were demonstrations of parachuting and ballooning, mock combats and slow flying races; ice cream stalls and Ascot hats by the score. But amidst all the excitement, three events really had the crowd on its toes. The first was No 43's 'tied together' aerobatics and the second was the synchronised pair from No 1 Squadron. *Flight* again:

'. . . this year having Furies they were able to do almost exactly as they wished, and the result was as near perfection as we are ever likely to see . . . their aerobatics were faultless and perfectly timed throughout'.

The correspondent goes on to describe the feats of the three-man team with similar enthusiasm:

'. . . In vain did we seek for any sort of flaw in the combined aerobatics by a flight of three Furies of No 1 Squadron. No 1 has not had Furies for very long, but there seems to be something in the air of Tangmere which makes for very high class flying. Bravo, No 1 (Fighter) Squadron'.

Over the next few years the rivalry of the two crack squadrons of Tangmere became legendary. First one and then the other would manage to pull off the honours at a host of displays, air defence exercises and gunnery camps. Their rivalry was intense, but it was full of good humour which extended from the Commanding Officers to the most newly arrived aircraftsmen and pilots. A proof of the esteem in which both squadrons were held is that nearly every graduate of the flying and trade training schools applied to be posted to one squadron or the other. Their mottoes, 'In Omnibus Princeps' and 'Gloria Finis', became reciprocally translated by the squadrons as 'First in the Bus' and 'The Arse End'. The standards of flying set by the Tangmere squadrons were so consistently high that they became the criterion by which others were judged. At this particular time, to be the best in the RAF meant that a squadron was almost certainly better than anything the world had to offer. 'In Omnibus Princeps', indeed!

By this time, No 25 Squadron had also joined the Fury set and No 3 Wing was once again equipped with the same type throughout. At the 1933 Hendon display the new boys were graciously permittd by No 43 to try their hands at 'tied together' aerobatics, but the show was virtually rained off. During the summer, No 1 pilots gave displays at Brooklands, Andover, Hatfield and Speke, increasing their reputation and making new friends wherever they went.

By the end of the year, most of the senior pilots of No 1 Squadron were posted. Sqn Ldr Spackman and his two Flight Commanders left within weeks of each other, to be replaced by Sqn Ldr Roy Chappell MC and Flt Lt Edward Russell-Stracey. As Ian Bertram had also recently left on his promotion there was now a distinct lack of experience at the top end of the Squadron. John Davys was therefore given acting rank to take temporary command of 'B' Flight. Sgt Philip Lee was promoted to Warrant Officer 2nd Class and became the first pilot of that rank to fly with the Squadron.

No 1 soon settled down under its new management, and during the winter several new pilots were trained to operational standard. When Flt Lt Noel Fresson was posted in to take over 'B' Flight, John Davys took charge of organising the coming season of air displays. At Hendon in June, Davys was entered for the solo aerobatic display, backed up by a synchronised pair composed of Sgt S. W. Bannister and A. Wroath. Davys' flight drew the now routine praise from the press and *Flight* wrote the following of the synchronised pair:

'As usual, the overture was breathtaking. The two Furies dived together, pulled up into a vertical climb, stalled outwards, half rolled to an inverted position and dived out. Then movement followed movement so rapidly that the eyes were dazzled. Rolls, rocket loops, upward rolls and slow rolls were mingled in perfect pattern to the organ notes of the two Kestrels'.

The highlight of the summer of 1934 was the visit to Canada by five of the Squadron's Furies, which were shipped across the Atlantic and reassembled at Hamilton. The party was led by Wg Cdr G. C. Pirie, MC, DFC, the Station Commander of Tangmere, and consisted of Flt Lt Russell-Stracey and Fg Offs Chatterton, Donaldson and Dixon. They had been invited across by the Canadian Government to perform an aerobatic display at the Canadian Centennial celebrations at Toronto on 2-4 July. This was a tremendous honour for the

Squadron because it was the first time that an RAF unit had been selected for a display in North America, and the Canadians had asked specifically for No 1.

The tour was a great success. In addition to the Toronto display, the Furies visited air shows throughout Ontario and Quebec for the following three weeks. They created enormous interest and admiration at each one, and the party returned to England at the end of August with a great sense of achievement.

The Air Ministry was now well aware of the dark forces stirring in Germany. Intelligence reports spoke of the secret air force that Adolf Hitler's Nazi Government was in the process of building up under the disguise of a large civil airline and a network of flying clubs. This growing awareness had not until now been reflected in the attitude of Parliament; in July 1934 the Secretary of State for Air had announced that Britain obviously had no need of an enormous fleet of aeroplanes and that there were no plans for expanding the RAF. It was left to the embarrassing Mr Winston Churchill to put the ugly truth before Parliament:

'Germany is arming secretly, illegally and rapidly . . . Besides their organised military machines they have a gigantic civil aviation with enormous resources of pilots and aeroplanes. A great number of these could be converted into bombing machines within a few days or even hours'.

The distant gathering storm had little immediate effect at Tangmere. Sqn Ldr Cedric Hill had taken over No 1 Squadron in September, and life continued much as before into the summer of 1935. Once again, the Squadron was well represented at air displays throughout the south of England, but the highlight of the year was the Silver Jubilee flypast.

On 6 July, 37 squadrons of the RAF were drawn up on parade at Mildenhall for inspection by King George, now celebrating the 25th year of his reign. No 1 Squadron had been there for some days, polishing its aircraft and buttons to a still greater brilliance. The 12 Furies were wheeled to their allotted places and the crews took post in front. The pilot of

Flt Lt Theodore McEvoy (white flying suit) with some of his pilots in 1936, shortly after the death of King George V (note the black armband). The old style of squadron badge can be seen on the fin.
Squadron Archives

Edward R I

J. D. Heaton-Armstrong

Chester Herald

and Inspector of Royal Air Force Badges.

College of Arms
July, 1936

The Hawker Fury was the fastest fighter of its era, and it was the first of five types of Hawker aeroplane designed by Sydney Camm to be flown by No 1 Squadron. This one carries on its fin the new style of squadron emblem adopted in 1936, after some argument with the Chester Herald.
Squadron Archives

each machine, dressed in spotless white overalls, was flanked by his rigger and fitter clad in their best uniform of pantaloons, puttees and the old dog-collar style tunic.

His Majesty spent well over an hour driving past his assembled squadrons before driving to Duxford, where he joined Queen Mary in watching 20 selected squadrons fly past the saluting base. The heavies came first and were followed by 11 squadrons of fighters. No 1 Squadron swept by at the head of 30 Furies of No 3 Wing. Finally came the new Gloster Gauntlets of No 19 (Fighter) Squadron which, after four proud years, had taken the blue riband of speed away from Tangmere.

Even as the red and silver Furies looped and rolled their way through the summer of 1935 with traditional elegance, Hitler's Italian allies were subduing the Abyssinians with aerial bombardment and poison gas. This news finally stirred Parliament into action, and a hastily improvised force was scratched together and sent to Egypt to protect Suez. Many of the best pilots were withdrawn from squadrons throughout the RAF to cope with the resultant expansion: No 43 Squadron was cut down to a mere cadre for many months, and by November, No 1 had only 15 pilots on its strength. No 43's loss was No 1's gain, when, as a result, Flt Lt Theodore McEvoy was transferred to the Squadron. He had commanded a Flight with the 'Fighting Cocks', and took command of No 1 when Sqn Ldr Hill was hurriedly detached to a staff post in January 1936. As the number of pilots was still diminishing, he held the command as a Flight Lieutenant for a full year. This closing of the ranks did not prevent the Squadron's full and immaculate participation at Hendon, although the display was somewhat reduced in size and dulled by the lack of competition from No 43. There was consolation, however, in the sight of two sleek, beautiful and very fast prototype monoplane fighters from the Supermarine and Hawker factories. The Spitfire and Hurricane had arrived.

During the summer it was announced that the present formal style of squadron badge was to be introduced for all RAF units: and squadrons were invited to tender designs of their choice to the Chester Herald for Royal approval. No 1 Squadron's existing emblem consisted of a pair of RFC wings sprouting from a wreathed roundel on which was superimposed a large black '1'. This apparently would not do. ACM Sir Theodore McEvoy tells the story:

'We were all very attached to our old 'Flying One' badge that had been used in World War 1, and we were highly incensed when the Chester Herald told us it was not "heraldic". I got Joubert, our AOC and a former squadron commander, to appeal personally to Heaton-Armstrong to let us keep the badge hallowed by twenty years' glorious use, but nothing less than 600 years cut much ice with the College of Arms, and we were invited to select a pair of "heraldic" wings from a set of samples sent us by the Chester Herald.

'Some of these were terrible — droopy things like dejected angels — but with Joubert's help we got him to devise the quite healthy pair that His Majesty eventually approved.

'Not long afterwards we were all busy with our paintpots putting the new badge on the fins of our Furies'.

As a result of a motor-cycle accident some years previously, McEvoy suffered from a very stiff back and when strapped into his cockpit he was unable to turn his head sufficiently to see what was going on behind. He therefore used to wear a small mirror strapped to each forearm to check up on the activities of the aircraft he was leading. The sight of one of the CO's arms raised in the air invariably resulted in the most immaculate formation flying from his pilots!

Life at Tangmere now continued against a winter background of growing crisis in Spain. The rumours of new high performance German warplanes were proved to be devastatingly true when Hitler threw his Condor Legion into the Spanish Civil War. When the swift and deadly Messerschmitt 109s began to tear the opposing biplanes to shreds, No 1 Squadron's pilots began to look at their beloved Furies in a new and sadder light.

The expansion scheme of the RAF was now accelerated by the flaunted ambitions of Germany. The new training schools began to produce their first graduates, and the best of these were posted to the fighter squadrons. By the end of 1936 Sqn Ldr Hill had returned to resume his command of the Squadron and 10 new pilots had arrived to bring the strength back to its former level. Flt Lt McEvoy was posted, and his place as Flight Commander was taken by Flt Lt Teddy Donaldson. No 1 was now very inexperienced. Apart from the CO and Donaldson, only Plt Off H. E. C. Boxer and some of the NCO pilots had spent more than a year on an operational unit. Among the new faces now arriving were the first of those who would be leading the Squadron into action in three years' time. Plt Off P. R. Walker, inevitably known as 'Johnny', and Acting Plt Off Prosser Hanks were in that batch. They and their compatriots soon impressed their mentors with their willingness to learn and their general zest for life. No 43 Squadron was also building up to full strength and, like No 1 Squadron, was getting a lot of good new blood. The outstanding recruit was Caesar Hull, who became a close friend of Prosser Hanks. Not long after their arrival these two took off in the Station Flight's two-seat Audax and proceeded to change cockpits in mid-air. To confound the disbelievers who accused them of landing in a field to effect the switch, they repeated the stunt over the airfield in full view of everybody. For this they were very quickly roasted by their respective squadron commanders, but the unspoken approval of their spirit was soon obvious to all.

Intelligence reports on the growth of the Luftwaffe were now becoming very ominous. It was quite apparent that if Britain found itself at war with Germany in the near future, the RAF would be hopelessly outnumbered and outclassed in everything save the quality of its men. Now all but the most complacent politicians were jolted into an awareness of the strength and ambition of the Axis powers. The country would have cause to be deeply grateful for the foresight of the Hawker Aircraft Co, which had laid down the initial production line of Hurricanes as a private venture without waiting for government orders. The volunteer reserve forces had been allocated to their war stations and in 1936 over 100 part-time airmen reported to Tangmere for their annual camp with No 1 Squadron. As the months passed, the Squadron quietly and efficiently continued its training. In a country at peace there was still time to enjoy life to its fullest. The coast, the countryside and the sky still beckoned as their playground, but since the crises of Abyssinia and Spain, things could never be quite the same. Tangmere's Golden Age had gone forever.

CHAPTER SIX

The Gathering Storm

By the spring of 1937 the expansion of the RAF was well under way. As the new eight-gun monoplane fighters were not yet ready for service, new squadrons were being formed to await them, temporarily equipped with biplanes. At the beginning of March it was learned that 'B' Flight of No 1 Squadron was to be detached permanently to re-activate No 72 Squadron, which was to be the first unit to get the Gloster Gladiator. The news was received with mixed feelings, for no one was keen to leave when rumour was rife that No 1 was on the short-list for re-equipment with Hawker Hurricanes.

Sqn Ldr Cedric Hill was promoted in May and, to the delight of his old squadron, he was appointed Station Commander of Tangmere. His place at the head of No 1 was taken by Sqn Ldr F. R. D. Swain AFC, who, during the previous September, had established a new world altitude record in the Bristol 138a monoplane. Dressed in a pressure suit that had made him look like a cross between Ned Kelly and a deep-sea diver, he had reached 49,967ft. He often remarked that he would have made it to 50,000ft had he forgone a second slice of toast at breakfast.

The Squadron, having lost 'B' Flight, was again very lacking in age and experience. H. E. C. Boxer was promoted to Flying Officer and became a Flight Commander in the temporary absence of Teddy Donaldson. The remaining pilots consisted of seven pilot officers and six sergeants, and among the new arrivals were Acting Plt Offs C. G. H. Crusoe and M. H. ('Hilly') Brown.' The latter, a rugged little Canadian, had crossed the Atlantic to join the RAF after reading an article on the growing strength of the Luftwaffe.

When Donaldson returned from No 72 Squadron in May, he sat down with the CO to organise an aerobatic team for the summer. It was decided that Donaldson was to lead and that the team should consist of four aircraft if suitable pilots could be found. After a week of practice they selected Fg Off Boxer and Plt Offs Walker and Hanks. These four, destined to become the most celebrated prewar team of all, flew together exclusively for the next fortnight, and were duly selected to fill the star billing at Hendon.

For Johnny Walker and Prosser Hanks this was a special achievement. Neither had seen much more than a year's total service and neither had over 200 hours in his logbook. Hanks, it must be added, was still a teenager.

In the Coronation Year of King George VI, 1937, Hendon was attended for the first time by a ruling monarch. The Air Pageant was as great a success as ever, with No 1 Squadron the first to mount a four-man aerobatic team. Donaldson's young team were the stars of the show and won a rare standing ovation from the crowd. For many of the public this final Hendon display would be the last opportunity of appraising the skill of the young men who were soon to stand between them and invasion.

After further practice, and following an aerobatic competition against other crack squadrons held at Northolt, No 1 was delighted to hear that its team had been selected to represent the RAF at the International Aviation Meeting at Zurich. This was a very great honour indeed, as it was the first time that the Air Ministry had authorised a RAF unit to represent Britain at a major international meeting on the continent.

The Zurich display was significant because the European powers were anxious to show the calibre of their pilots and equipment to the worried world. It was plain that the Germans and Italians were going to be there in force. A British success was therefore all the more desirable in terms of prestige.

The six Furies selected for Zurich included the four normally flown by the team members. All six were given a thorough overhaul during July with help from the Rolls-Royce and Hawker representatives at Tangmere. While K2881, K2043, K5673 and K2089 were receiving such loving attention in the hangar, their four pilots flew together in other machines several times each day to perfect their sequence. Donaldson's insistence on constant practice whatever the weather was to pay dividends in Switzerland.

The Zurich team in 1937. Left to right: Plt Off Prosser Hanks, Plt Off Johnny Walker, Fg Off 'Tops' Boxer and Flt Lt Teddy Donaldson. Their display, in torrential rain, stole the show from the Luftwaffe's solo pilot and Italy's crack squadron of 'stunting bachelors'. *Squadron Archives*

'Tops' Boxer flew in the No 2 position on Donaldson's starboard wing, while Johnny Walker held the similar No 3 position to port. Hanks flew line astern on his leader in the new and somewhat controversial 'box' position. Some mystique had been woven by the press about the problems of flying in the box, which is, if anything, slightly easier than in the other two echelon positions. Young Prosser Hanks made it seem quite effortless.

All was eventually ready. The groundcrew and spares set off for Zurich in a Bristol Bombay, to be followed by the six Furies led en route by Wg Cdr Cedric Hill in one of the spare aircraft. They arrived at Zurich's Dubendorf airfield on Saturday, 25 July, just in time for a practice before the show opened on the following day.

It was estimated that almost half of Zurich's population turned up at Dubendorf on the first Sunday. The team's display drew tremendous admiration from the crowd and the international press. Throughout the week the team continued to match the finest that Europe could produce, drawing considerable open praise in the process from their rivals. On the Monday they were visited by General Milch, then Chief of Staff of the Luftwaffe, who told Donaldson that he had never in his life seen flying like it.

As the week progressed towards the climax of the final Sunday, it was noticeable that the opposition was becoming increasingly daring in its efforts to outshine No 1 Squadron. Italy's crack squadron of 'stunting bachelors', who normally put up an excellent show in their Fiat CR42s, were getting visibly bolder as the weekend approached.

There was no display on the Friday as the airfield was being readied for the weekend. Most participants took the day off. The team had been given a car to take them to Lucerne for the day, but did not go until Donaldson had led them through their routine once more at breakfast time. This amazed their overhung rivals who had last seen the team whooping it up at a party in the small hours.

The weather was fine for Saturday's display, when the Furies took equal honours with the Italian team. The Fiats put on a most exotic show which pleased the crowd with a lot of low flying, sideslipping and noise. The German solo artist was also impressive. He began his display with a loop that led into an inverted dive, from which he bunted with his fin about 10ft off the grass.

The scene was now set for a lively and competitive final display on the Sunday, but the weather was threatened by thunderclouds lurking ominously in the hills. The storm held off for the first part of the show but then the skies opened and the team took off in a tremendous cloudburst. At times the cloudbase was down to 200ft. Donaldson somehow threaded the team in and out of the bad weather, looping and rolling as he got the opportunity, and in spite of the rain blinding their goggles and windscreens, they completed the entire sequence. The Furies landed to tumultuous applause that was given to no other act. The rest of the programme was then abandoned because of the weather. This must have been particularly galling for the Italians, who had asked to delay their appearance to provide a grand finale, but the Fiat pilots were the first to offer their warm congratulations to No 1's team.

The party returned, via a memorable night in

Paris, to a tremendous reception at Tangmere and the adulation of the British press. For a public worried by the growing strength of the Luftwaffe, here was proof that at least one RAF squadron could take on the best and fly the pants off them.

Rumours of impending German aggression grew stronger throughout the winter, but there was still no sign of the Squadron being re-equipped with modern aircraft. Six years previously the Fury had been the fastest fighter in the world, but by 1938 it was the slowest in the RAF, with no hope of intercepting modern bombers. The pilots had discovered in exercises with the new Bristol Blenheim that only a head-on attack carried any hope of success,and the chances of being in position for one were virtually nil.

The prospect of a German invasion called for a lot of strafing practice. This involved diving the Fury steeply at 260mph from 1,000ft at ta gets set up on the coast. Sgt C. G. Tucker struck the ground while pulling out from such an attack on 2 November, severely damaging his undercarriage, and choosing to risk a forced landing rather than bale out, he orbited Tangmere for 90min to use up his fuel. The Fury tipped on to its nose during the ensuing landing, but was flying again two weeks later.

When Sqn Ldr I. A. Bertram assumed command of No 1 Squadron in January 1938 he inherited a team of very determined young pilots. War with Germany was now becoming an increasing possibility, and no one could know which would come first, conflict or the Hurricane. Sqn Ldr Bertram had been a Flight Commander with the Squadron in the early days of the Fury and was delighted to see the way his pilots now handled it and were developing new ideas to cope with the disadvantages of its performance. The rigid squadron attack procedures had changed little over the years, and were soon to be proved too inflexible for the higher speeds of monoplane combat.

New pilots arriving on the Squadron were subjected to an intensive course in air combat. Whatever their background, they were left in little doubt of their lowly place in the scheme of things. It took several months for a new boy to win an accepted place in the bosom of the unit, and it was soon obvious to all if one of them was not making the grade. The new pilots now arriving were a very cosmopolitan crew. Some, like Sgt Bill Berry from No 43, were seasoned regulars, but they were being joined in increasing numbers by volunteers from the wider world. Plt Off Billy Drake from London, Leslie Clisby from Adelaide and 'Pussy' Palmer from Cleveland, Ohio, were among the new faces of 1938. It was up to those at the top end of the Squadron to knock off their rough edges and mould them into the collective image without dampening their enthusiasm or killing their spirit. This would now have to be done without the help of Flt Lt Teddy Donaldson, who had been posted to Hinaidi — it was hard luck on the former team leader that he should be sent to one of the few unaccompanied stations in the RAF very soon after his marriage. He went on to fight with considerable gallantry with the Gladiators in Norway, and later took the World Air Speed Record in a Meteor. After the war he retired from the Service to become the Air Correspondent of the *Daily Telegraph*.

ACs Alf Tupper, Boz Bosworthick and George Millman discuss the recovery of a bent Fury. FS Gillie Potter and Sqn Ldr Bertie Bertram (above the propeller) look on in disgust. *E. Bosworthick*

The Sudetenland crisis of that summer led to the visit of Mr Chamberlain to Munich in September, and the British services began hurriedly to prepare for war. The effect of the 'flap' at Tangmere was quite startling. The first thing to go was the envied tropical routine, which brought the station into line with rest of the RAF in England. Air raid shelters began to appear beside the hangars and throughout the domestic site. All leave was cancelled. A critical shortage of belted ammunition for the Vickers guns of the Furies was discovered, but this was quickly remedied by shifts of officers and men who spent days and nights tapping rounds into the belts with large spanners, because no belt positioning machines were available — even those pilots detailed to stand

After Munich: the Squadron's Furies, now camouflaged, are lined up tail-to-tail for a final photograph. The first Hurricanes arrived soon afterwards. *Squadron Archives*

by at dawn worked late into the evenings on this thankless task. The hangars were camouflaged against air attack, and guards were mounted on the aircraft by day and night. Then came the final injustice — the order to camouflage the Furies. The years of loving polish, of bright red stripes and of black and white chequerboards, all had to go in one long orgy of spray and paintbrush, of dark greens and browns.

Sqn Ldr Bertram called his pilots together for a final briefing. As the Fury did not have the speed to intercept the enemy, he told them, head-on attacks would be obligatory. If the ageing Vickers guns failed to destroy the target, it was to be rammed. A very sobered collection of young men left the briefing room.

The Prime Minister returned from Munich with an assurance of no further German expansion: Britain would not be going to war that year, and the nation began to breathe again. For the RAF, Munich provided the time for its vital re-equipment and gave another year of life to its young men. Chamberlain's act of appeasement, so often reviled, probably saved Britain from a terrible defeat.

The intensive flying programme for No 1 Squadron lasted throughout the winter and beyond. The day of re-equipment could not be far away and the CO intended to get every last ounce of value out of the Fury before the Hurricane arrived. Practice dogfights with No 43 were all part of the routine, while the Ansons of No 217 Squadron, also at Tangmere, were frequently used as interception targets. Even these relatively slow aircraft emphasised the dated performance of the Fury.

Then, one day in early November 1938, a hump-backed little fighter burst into the Tangmere circuit accompanied by the unforgettable growl of its Rolls-Royce Merlin. The Hurricane had come to No 1. To the obvious envy of the 'Chicken Farmers' the Squadron began to receive its new aircraft in twos and threes over the winter. With considerable nostalgia the Furies were lined up, tail to tail in accordance with tradition, for a final photograph. It is sad to relate that not one

A new Hurricane being refuelled at Tangmere early in 1939. Note the original fixed wooden two-bladed propeller. *Squadron Archives*

example of this most beautiful of biplanes has survived.

By March 1939 all the pilots of No 1 Squadron had been checked out to fly the Hurricane. To their surprise they found that the new aircraft handled very much like the Fury, for its controls were perfectly balanced in the true Hawker tradition. It took them some time to get used to the speed of the Hurricane, but after a few hours in the air they found that their reactions were getting noticeably faster.

As with every new aircraft there were teething troubles. The early Hurricanes were equipped with a wooden fixed-pitch propeller which could not take full advantage of the tremendous power of the Merlin engine, and the rate of climb of the early production Hurricane was little better than that of the Fury. The aircraft was armed with eight 0.303in Brownings, but in the early months the guns were prevented by gunbay icing from delivering their punch above 15,000ft.

The press gave the Hurricane a lavish build-up. It was described as being 'sleek and venomous' and 'the 400mph miracle fighter plane' — that 400mph tag was a myth that stemmed from a single proving flight from Edinburgh to London on the wings of a 90kt gale. The true top speed of the aircraft was nearer 320mph.

One newspaper ran a story about the new breed of supermen who were being specially developed by the RAF to fly its new fighters. These pilots, it was claimed, were abstemious of all vices. Their biceps bulged and their hands were gnarled with wrestling with the controls of their Hurricanes. This news was received at Tangmere with the hoots of mirth that it deserved. Abstemious indeed! The publicans and fathers of Sussex had good cause to know otherwise.

In April, soon after conversion to the Hurricane was completed, the 'Bull' arrived at Tangmere. Sqn Ldr P. J. H. Halahan was well nicknamed. The stocky Irishman who was to lead No 1 Squadron to war was considerably older than his pilots, with an imposing manner and a voice to match. To take command of No 1 he returned from the Middle East where he had been attached to the Trans-Jordan Frontier

Force and had served on the staff of HQ Iraq. He was an old Cranwellian and had served for a while with the Fleet Air Arm.

As the new CO had not flown very often during his previous tour, he arranged a course of dual instruction for himself with his friend George Stainforth, a former Schneider Trophy pilot, who was writing the handling notes for all the modern types of aircraft. Before long he was ready to fly the Hurricane and soon reassured his pilots of his ability to lead the Squadron in the air.

The days of early summer started to pass all too quickly. It was becoming obvious that the Munich agreement had only provided a stay of execution on the long, easy years of England's peace. At Tangmere the training programme was intensified, and No 1 Squadron flew up to Sutton Bridge for its air firing practice camp at the end of May.

At Sutton Bridge the pilots had their first opportunity of firing the eight guns of their Hurricanes. The initial results were very discouraging. On the orders of ACM H. C. T. Dowding, C-in-C Fighter Command, the guns of the new monoplane fighters had been harmonised to give a widely spaced bullet pattern at 400yd range. At the time it was thought that the short range German fighters would never penetrate as far as England, and this bullet pattern was designed to destroy the enemy bombers which would present a much larger target.

But Sqn Ldr Halahan was very opposed to this idea. Any target, he argued, would be more vulnerable to a concentrated and accurate burst of fire. Dowding's bullet pattern, known as the 'Stuffy Spread', certainly stood a greater chance of simply hitting the target, but the CO considered that the armour of the German bombers would probably save them from total destruction. He would have preferred to harmonise the guns on a single point 250yd in front of the aircraft, but the order to spread their field of fire was most rigidly enforced.

On his first firing sortie, Bull Halahan registered precisely four hits on the towed target. Johnny Walker flew next and missed completely, while Prosser Hanks got the highest score with five. As all three pilots were first-class shots, it was obvious that the 'Stuffy Spread' was just not good enough. With the co-operation of Wg Cdr Soden, the CO of Sutton Bridge and a former No 1 Squadron Flight Commander, Halahan had the guns of one Hurricane spot-harmonised to 250yd. On the next sortie he blew the target to pieces with his first burst. At the risk of a

Hurricane Is taxying out for the Empire Air Day flypast at Tangmere in July 1939. No 1's original code letters (NA) were removed in France.
Wg Cdr P. J. H. Halahan

court-martial the CO had the remaining aircraft secretly reharmonised before they returned to Tangmere.

During the spring the Air Ministry had announced a scheme to affiliate the more famous RAF squadrons to towns and cities near their peacetime stations. No 1 became 'Brighton's Own', and a link was forged that has happily survived the war and all the subsequent moves of the Squadron. The citizens of Brighton took their first opportunity of visiting their own fighter squadron in June, when a large crowd turned up at the annual Empire Air Day at Tangmere. They were particularly impressed when Sqn Ldr Halahan led 15 Hurricanes on a low diving pass of the airfield in tight formation. The Squadron visited Brighton on 8 July with a flypast and mock combat over a Mayoral Garden Party at the Brighton Pavilion, the display culminating in a noisy beat-up of the seafront that delighted the Mayor and most of the population.

Throughout the summer the training programme continued by day and night. Exercises were flown in co-operation with the new and highly classified RDF (radar) stations; although the initial results were fairly encouraging, it was some time before workable procedures were developed.

The introduction of searchlight batteries throughout the south of England resulted in a considerable amount of night flying affiliation exercises for the Squadron. Night flying has always been unpopular with day fighter pilots, and Flt Lt Willy Hemmings was killed when he became disorientated, dived down a searchlight beam and hit the ground hear Goodwood. This was a tragic loss for the Squadron because Hemmings had been an efficient and popular Flight Commander. The CO had Johnny Walker promoted to take his place.

An incident with a slightly happier outcome befell Prosser Hanks in July. His engine cut out over Hayling Island while he was co-operating with Naval ack-ack and he had to make a forced landing on the crowded beach at Sandy Point. Despite every effort to avoid the holidaymakers, he struck and slightly injured a disabled child. The boy's parents later expressed their appreciation of the pilot's efforts to avoid disaster, and Prosser got a sympathetic write-up in the *Daily Mirror*. The paper described his 'desperate wrestling with the controls' in heroic terms.

The annual Air Exercises were held in August, during which a foreign air force was allowed to fly over England for the first time,

Sqn Ldr Bull Halahan, Flt Lt Johnny Walker and Bill, the Squadron's mascot. *Wg Cdr P. J. H. Halahan*

Plt Off Paul Richey kept a detailed diary of his time with No 1 in 1939 and 1940. This was later published as the classic book *Fighter Pilot*, which tells the story of the Squadron's exploits in France. *IWM*

when French bombers made a mock attack on London. Paul Richey describes the event in his book *Fighter Pilot*.

'We intercepted those bombers, with the aid of RDF, at 14,000ft in bright sunshine off the South Coast. How gay they looked in their red, white and light blue markings! But how pathetically out of date . . . Later that day, Johnny Walker and I intercepted three RAF Blenheims low over the Downs underneath a big thunderstorm. They looked grey and grim — and they had mock German crosses painted on them in white.'

After the exercises the Tangmere squadrons gathered in a briefing room to hear a lecture on the current strength of the Luftwaffe from Sqn Ldr Coop, the British Assistant Air Attache in Berlin. It seemed that even the wildest rumours had understated the size of the German air force, and the pilots listened in stunned amazement. Grim-faced and determined, they now began to realise the immensity of the task that, with deadly certainty, now lay ahead.

Arrangements had been made between the British and French Governments which resulted in the formation of the British Expeditionary Force that was to proceed to France on the outbreak of war. Fighter cover for the army was to be provided initially by four squadrons of Hurricanes, and Nos 1, 73, 85 and 87 were selected.

The decision to send four front-line fighter squadrons to France was bitterly opposed by ACM Dowding. He even paid them the compliment of saying that he could not hold himself responsible for the defence of London without them.

Within a few days of mobilisation over 100 members of the RAF Volunteer Reserve had reported to Tangmere to serve with No 1 Squadron. The Squadron was fully established with pilots, riggers and fitters, so the new men were given the responsibility for the unit's mobility and upkeep in the field. Soon after their arrival, Sqn Ldr Halahan assembled them in the hangar for a talk on the Squadron's traditions of teamwork and on the high standards of conduct, discipline and professionalism that would be required if No 1 was to survive against the Luftwaffe. He allocated a group of reservists to each of the regular airmen for guidance and tuition, and before long they became an integrated and indispensable part of the Squadron.

None of the reservists bore NCO rank, which proved awkward as they were to be formed into complete subordinate sections of the Squadron. Some of them obviously had to be promoted. Sqn Ldr Halahan solved the problem with a commendably democratic gesture by asking them to elect their own NCOs!

A few days later, those reservists selected as drivers were sent off to a vehicle park at Wembley to collect the various trucks and cars that would be required in France. They pillaged the establishment with great enthusiasm and returned with four staff cars instead of two, in addition to a complete mobile workshop to which the Squadron was not entitled. The CO was delighted by their initiative and made sure that their prizes were kept well hidden until the day of departure. From that moment on the reservists were fully accepted by the regulars as established members of No 1.

It was indeed fortunate that the old and new members of the unit settled down to work together so well. In the Squadron, where a nod from the CO was as good as a page of orders, there was no place for the blind disciplines of the parade ground, and still less for the passenger or the incompetent. In France they would have to become masters of improvisation: the key to their survival would be the enthusiasm, initiative and spirit of every man in the team. Over the months to come they would show the world that they held that key, but the odds against that survival were obviously very great.

On 1 September the Squadron was visited by 'Stuffy' Dowding. After a long and searching inspection of the aircraft and crews, he turned to Halahan and said:

'I am very sorry indeed to lose No 1 Squadron. If war breaks out and you go to France, you may only take 12 aircraft and the pilots you have. You can expect no reinforcement from me.'

On the morning of Sunday, 3 September, the officers of No 1 Squadron gathered around the Mess radio to hear the sad voice of the Prime Minister, who had striven so hard for peace, break the inevitable news to the nation. It was greeted with a long and thoughtful silence. That night Sqn Ldr Halahan completed his regular report on the Squadron's peacetime training activities. He completed it with the following sentence:

'If war comes tomorrow, no Commanding Officer could wish for a better squadron'.

CHAPTER SEVEN

The Phoney War

Since 30 August, No 1 Squadron had been standing by at one hour's notice to go to France. Thirty tons of stores lay crated in the back of the lorries and the 12 Hurricanes stood ready for the call to arms.

It had been predicted that swarms of German bombers would cross the coast on, if not before, the declaration of war, and both Nos 1 and 43 Squadrons were ordered to maintain a section at cockpit readiness. That night, the RDF station at Ventnor picked up a plot in mid-Channel. Flt Lt Walker was scrambled with Plt Off Richey and Sgt Soper to intercept; it was a false alarm, and they searched for an hour in the empty sky. Other sections were scrambled later and returned with nothing to report except that the blackout had taken effect all along the South Coast.

Throughout the week the Squadron waited for both the Luftwaffe and the move to France. On Monday the advance ground party left by rail for an unknown destination. In the absence of further news the Hurricanes were dispersed around the airfield and tents were pitched for the crews on standby. The uncertainty of the future prompted the Adjutant to pack up the Squadron's valuables for safekeeping, and the silver and a few trophies, such as the fins of two German aircraft shot down in 1918, were duly dispatched for storage. Their presence in the museum at Wittering today testifies to the wisdom of his action.

On Thursday the order arrived: the destination was to be Le Havre. The main convoy of trucks and ground equipment set off for Southampton without delay with Plt Off Blomeley in charge. At midday a signal arrived to say that the advance party had reached Octeville, the airfield that No 1 Squadron was to occupy just outside the French port. Little co-operation was forthcoming from the French authorities, and the advance party was having great difficulty in preparing for the Squadron's arrival. Sqn Ldr Halahan was never a man to accept such treatment of his Squadron, so he jumped into his Hurricane and flew over to see what the problem was.

After breakfast on Friday the pilots were visited at Tangmere by the Duke of Norfolk, who was organising the local Territorials, and by Col Richey, who had dropped in to see his son Paul. Soon after the visitors arrived the Squadron was called to readiness and ordered to take off for Octeville at 11.30. With apologies to their distinguished guests the pilots dashed around collecting their kit and stowing it in the Hurricanes. After a final photograph they removed the Squadron badges from their overalls and strapped in. One by one the Merlins coughed into life and Johnny Walker led the Squadron past Tangmere in a final salute. No 1 was on is way to war.

After arriving over Le Havre in the noisy Squadron tradition, they landed at Octeville to find their accommodation problem solved. Apparently the 'Bull' had marched up to the French Air Force authorities on the previous day and given them the sharp side of his tongue in schoolboy French. This had the desired results and he was directed to an adjacent nunnery which was being evacuated that night. His knock was answered by a novice nun who fled babbling at the sight of the fearsome Irishman. The situation was happily resolved by a charming Mother Superior who assured him that the place

would be completely nunless by evening.

That first night in France was spent exploring the possibilities of Le Havre. At La Grosse Tonne the pilots ran into the staff of the American Hospital, who had been evacuated from Paris. The nurses had seen and admired their arrival earlier in the day and were promised a full show for the morrow. An enjoyable pub-crawl followed, before the pilots wended their unsteady ways back to the nunnery and laid their heads in the tiny cells that had never before heard the snores of men.

The 'Bull', seeing the bleary eyes of his pilots in the morning, set them to work with the digging of slit trenches before leading them up for an hour of practice combat. On the way home they gave Le Havre a stupendous beat-up at the request of the British Consul to cheer up the departing American civilians, who were very worried by the political situation. Later in the day they paused from further trench digging to watch the arrival of No 73 Squadron from Digby.

***Fair Stands the Wind for France*. This painting by Chris Golds captures the moment as Johnny Walker leads the Hurricanes away to the war.**

At 18.00 they were ordered to move immediately to Cherbourg to cover the arrival of units of the British Expeditionary Force. The CO was reluctant to move as they had no maps whatsoever of the French coast, and it was not long before nightfall. But they were on the ground at the new airfield, refuelled and ready to go at 19.10hrs, feeling fairly pleased with themselves at such a slick move. No 73 had the last laugh by descending on Le Havre that night and drinking all the free beer that had been lined up for No 1.

Returning to Octeville on the 11th, they welcomed Plt Offs Matthews and Birch to the fold. Fg Off Harris also arrived to take up the duties of a full-time adjutant, much to the relief of 'Leak' Crusoe.

There was little to do for the next 10 days but air-test the guns and radios, fly over to inspect other airfields in Normandy, or scramble down the cliffs for a bathe in the Channel. The transport convoy arrived on 15 September. According to Plt Off Blomeley, they had had quite a journey because instead of being disembarked at Le Harve they had been set down at St Nazaire. After many adventures they had passed through Paris completely penniless, and Blomeley had called on the British Ambassador to draw money for food and petrol.

On the 25th, No 1 Squadron was ordered to move to Norrent-Fontes, near St Omer in the Pas-de-Calais. The MT convoy and advance party left immediately, and Sqn Ldr Halahan led the Hurricanes to their new airfield on the 29th, leaving their friends in Le Havre to the tender mercies of No 73. After the pleaseant surroundings of Octeville, Norrent-Fontes came as a rude awakening. The accommodation was under canvas and the falling rain that lasted for the next fortnight made their surroundings a far cry from the comforts of Tangmere.

At this stage it would be as well to explain the structure of the RAF Command in France. The force comprised the Advanced Air Striking Force and the Air Component of the BEF. The AASF, commanded by AVM P. H. L. Playfair at Reims, consisted of 10 Blenheim and Battle light bomber squadrons, while the Air Component, under AVM C. H. Blount, included the four Hurricane squadrons presently stationed on the Channel. Both came under the command of AM 'Ugly' Barrett as C-in-C BAFF (British Air Forces in France).

The C-in-C had been worried from the outset by the lack of fighter cover for his bombers in support of the French Army in the Maginot Line. His fears proved justified when a formation of Fairey Battles was badly savaged by Bf109s over the frontier. The survivors had requested fighter cover for future sorties and, as the French seemed neither capable nor interested in providing it, Barrett was obliged to transfer his two best Hurricane squadrons to the AASF.

No 1 had hardly finished digging in at Norrent-Fontes when the change of plan came through. Although this meant that the convoy and advance party had to move out again at dawn on 2 October, the news caused little dismay. The Squadron had decided not to like Norrent-Fontes.

To give the advance party time to reach the Squadron's new airfield at Vassincourt and prepare it for the arrival of the Hurricanes, No 1 stayed at Norrent-Fontes for a further week. During the week the CO flew east with Clisby and Brown to inspect the new airfield, and dropped in to interview the attached Battle crews at Ecury on the way home. The AOC came to Norrent on the 5th to give the pilots a rundown on the current situation; he remarked, as Dowding had, how sorry he was to be losing No 1 to the Advanced Air Striking Force.

The Hurricanes flew to Vassincourt on 9 October without incident, but the groundcrew were not so lucky. They were travelling in four French Air Force Wibault transports and were fired on by British ack-ack over Arras, which mistook them for the similar Junkers Ju 52/3m troopcarriers. One was damaged and had to return to Norrent for repairs. Sqn Ldr Halahan, hearing the news at Vassincourt, immediately rang BEF Headquarters and gave the 'pongos' a fearful blast on the qualities of their aircraft recognition and parenthood.

Vassincourt lay in pleasant wooded countryside some 50 miles east of the Air Headquarters at Reims. Nearby was the provincial town of Bar-le-Duc, and the airfield was situated on a hill above the village of Neuville, which the groundcrew promptly pronounced to be 'a dump'. The advance party had selected some warm and dry stables for the airmen and had acquired a malodorous and tumbledown house for the Officers' Mess, whose members were to be billeted with individual families in the village. The young pilots took one look at their intended Mess and offered to do an immediate swap with the airmen, but the CO had a better idea. Grabbing Lt Moses Demozay, the Squadron's French Air Force interpreter, by the arm, he marched off to see the mayor and bribed him to move out of the town hall. M le Maire, knowing on which side his bread was buttered, shifted across the road to the old ruin refused by the officers with only a token protest. Like the rest of the village, he was delighted at the thought of the extra trade that Les Anglais would bring. To prove its good faith, the Squadron put on a display for the villagers, and local relations were henceforth excellent.

No 73 Squadron had moved eastward at the same time and was now based at Rouvres airfield, some 20 miles to the north, near Etain. Wg Cdr Walter and Sqn Ldr D. A. Pemberton had formed No 67 Wing Headquarters at Bussy-la-Cote, a village two miles from Neuville, and had laid telephone lines to the Squadron.

Plt Off Boy Mould — the first RAF pilot to shoot down a German aircraft in World War 2. *IWM*

Seventy miles to the northeast lay the massive ramparts of the Maginot Line, and beyond them — Germany. Until now, the front had been very quiet, for the Luftwaffe, still busy in Poland, had limited its activities over France to occasional sorties by high-level reconnaissance aircraft. These were mostly Dornier 17s and Heinkel 111s which had occasionally been spotted passing high above Vassincourt.

For the first few days the Squadron's flying was devoted to sector recces of the area up to the frontier. This gave the pilots a chance to get to know the local landscape and diversion airfields. On the 12th, Prosser Hanks took a Hurricane to Reims for some combat trials with French Curtiss and Morane fighters, whose performance he found to be definitely inferior.

No 1 Squadron's first real operational task of the war was a frontier patrol which took place on 15 October. Led by their Commanding Officer, 12 Hurricanes refuelled at Etain before spending almost an hour over the lines. They penetrated as far as Saarbrucken where heavy and accurate flak was encountered, but no enemy aircraft were seen.

The precise role of No 67 Wing was clarified on 22 October, at a conference at Reims attended by Wg Cdr Walter, Sqn Ldr Halahan and the CO of No 73, Sqn Ldr 'Red' Knox. The primary task of both squadrons was to escort bombing sorties in the event of the expected German invasion of France. While waiting for Hitler to strike, they were also to fly standing patrols in the hope of destroying the high-level German reconnaissance aircraft. Accordingly, three likely patrol lines were fixed. No 1 was to be responsible for the 'right front' between Nancy and Metz, while 73 was to look after the 'left front' between Thionville and Verdun. Either squadron could be called upon to reinforce the other's area, and the 'centre court' would be patrolled by both units on an opportunity basis. In addition, each squadron was to maintain a section at cockpit readiness in good weather for aerodrome defence.

The opportunity for action came on 31 October. At 14.30hrs three unidentified aircraft were spotted over Vassincourt at 20,000ft and the aerodrome defence section was scrambled to intercept. As the three possible raiders had split up, the Hurricanes fanned out in pursuit with boost override plugs pulled to achieve the fastest possible climb. The only pilot to make contact was Plt Off 'Boy' Mould, who identified his target as a Dornier 17. He attacked from astern and after a long burst of fire the German aircraft fell in flames. It exploded on striking the ground, killing the entire crew.

Mould's Dornier was the first aircraft to be shot down by the RAF in France since 1918. A great deal of publicity resulted from the kill, and the Squadron had to suffer a flood of war correspondents from Reims, all thirsting for a good story for their readers. Their subsequent visits came to be anticipated with gloom, for most of the reporters proved to be very tiresome and incredibly ignorant of their subject matter. The CO was very much against individual publicity for his pilots, in case it jeopardised the Squadron's teamwork. In this he had the full support of his men who were quite relieved when the 'Bull' banned all reporters from entering the Officers Mess.

Ten miles from Vassincourt, at Bar-le-Duc, No 1 Squadron found a home under the hospitable roof of the Hotel de Metz. This friendly provincial hotel was run by the Ensminger family, who quickly took the boys to their hearts. They organised mending and washing, and also prepared vast meals at short notice. Jean Ensminger was a superb chef, and his Breton wife — known to all as Madame Jean — always referred to No 1 as 'our squadron'. Together they provided a welcome haven of hospitality that was to work wonders for morale during the long winter that lay ahead.

Sqn Ldr Halahan was determined to take full advantage of the absence of the Luftwaffe to adapt tactics in the light of changing conditions. The Squadron flew frequent exercises with the Battles of Nos 15 and 78 Squadrons to perfect methods of attacking and escorting bomber formations and sometimes met No 73 for a big practice dogfight. As well as keeping the more experienced pilots on their toes, this practice was invaluable for training the more recent arrivals who seemed to have been rushed through training, and still had a lot to learn. One new pilot who needed little in the way of improvement was Fg Off John Ignatius Kilmartin — 'Iggy' or 'Killy' for short — who arrived from No 43 in early November. As a former 'Fighting Cock' from Tangmere he needed no introduction to No 1, and added some valuable experience to the Squadron.

The German aircraft over France until now had proved very difficult to see from below as their bellies were finished in duck egg blue. In contrast, the Hurricanes were painted black and white underneath as an aid to recognition by Allied anti-aircraft gunners. The Squadron had already been fired at by friendly ack-ack often enough to know that this was asking for trouble. The CO accordingly arranged for his aircraft to be painted underneath in the same light blue as the enemy. After a long argument with HQ AASF and the Air Ministry, the Hurricanes were allowed to remain inconspicuous. Later, this colour scheme was adopted as the standard finish on all British day fighters.

A dramatic improvement in the weather on 23 November brought the German reconnaissance machines over in force. At 11.45 Flt Lt Plinston was leading a patrol of three 'B' Flight aircraft between Verdun and Metz when he spotted a Heinkel 111 heading back into Germany. Overtaking it at the frontier, the section delivered a series of attacks and shot it down near Saarbrucken. Sgt 'Darkie' Clowes was delivering the mortal blow when six French Moranes appeared on the scene. One of them, desperate to get in on the act, came boring in and collided with the Hurricane's tail, removing one elevator and most of the rudder. Clowes put up an excellent show by

The pilots outside the Mairie at Neuville. Left to right: Drake, Clisby, Lorimer, Hanks, Mould, Sqn Ldr Halahan, Moses Demozay, Walker, Doc Brown, Richey, Kilmartin, Mitchell, Palmer.

Fg Off 'Pussy' Palmer from Cleveland, Ohio, who shot down a Dornier flown by Arno Frankenburger on 23 November 1939. *IWM*

Sgt Arthur 'Darkie' Clowes, looking rueful alongside his bent Hurricane. His tail had been chewed up by a French Morane as they both tried to shoot down a Heinkel, and he executed a most skilful forced landing. *R. A. Albonico*

Below:
Bull Halahan shows Frankenburger's gun to Richey, Clowes and Walker. The gun is still in the Squadron's museum. *Squadron Archives*

getting back to the airfield but had to land at 120mph to maintain control, and tipped up on his nose in the process. He was shaken, but fortunately unhurt.

Just after lunch, Bull Halahan and Hilly Brown were patrolling the same line when bursts of ack-ack drew their attention to a Dornier 17. They attacked individually from astern and sent their quarry down in flames. There were no survivors.

The third kill of the day was provided by a section of 'A' Flight led by 'Pussy' Palmer, which took off to intercept another Dornier seen passing high above Vassincourt. They caught it when it turned towards Reims, and Palmer led a stern attack which set one engine on fire and caused two of the crew to bale out. He then flew alongside to admire his handiwork and saw the pilot slumped over the controls, apparently dead. To his horror, the Dornier throttled back and swung in on his tail. The German pilot left his seat, grabbed a free gun and fired a burst into Palmer's aircraft before he too took to his parachute. Pussy ducked as he heard bullets pierce the back of his head-rest. One of them shattered his windscreen and another punctured his glycol tank, causing the engine to seize. He pulled off a most commendable forced landing in a field.

The story has an interesting sequel. The pilots later considered that the German had put up a very spirited fight, and on the following day had him removed from custody to dine with them at the Mess at Neuville. He identified himself as Unteroffizier Arno Frankenberger; once the ice was broken he turned out to be a most likeable and entertaining guest, with a background very similar to their own. He presented the Squadron with a photograph of himself for the diary, and was given a greatcoat and sweater before being marched back to the PoW camp.

Halahan later regretted that he had allowed Frankenberger to be entertained in the Mess, because the pilots liked him so much that they began to wonder whether the Nazis were as unpleasant as they had been made out to be. However, the Squadron did benefit in one way from Frankenberger's visit. During dinner he had revealed that the Luftwaffe went on leave each Friday, and were not required to fly again until Monday afternoon. Bull Halahan decided to check on this. An experimental mobile RDF set had just arrived at Vassincourt under the command of Fg Off Toby Carter, so for the next few weekends the radar set was employed to scan the skies above the Maginot Line. Absolutely nothing was seen, and the Squadron came to the conclusion that Frankenberger was right. As a result of this discovery it became possible to release people to Paris at regular intervals.

As a result of Fg Off Palmer's lucky escape with the Dornier, the CO managed to scrounge some armour plate from a wrecked Fairey Battle and had it fitted behind the seat of one of the Hurricanes. His experiment was opposed by higher authority, which assured the Squadron that it would upset the trim of the aircraft. Hilly Brown accordingly went to Farnborough with the modified aircraft, and put on a sizzling show of aerobatics to disprove this. Back armour was thereafter fitted throughout Fighter Command, and saved the lives of countless pilots — a direct result of Sqn Ldr Halahan's initiative.

Unteroffizier Arno Frankenburger, who shot down 'Pussy' Palmer before baling out and being taken prisoner. He was liberated for the evening by the pilots, and dined with them at Neuville.
Squadron Archives

Prosser Hanks landing at Vassincourt after a patrol.

Flt Lt Plinston was posted back to England at the beginning of December, and Prosser Hanks was promoted in his place to command 'B' Flight at the age of 22. In his three years with No 1 Squadron, Hanks had developed into a very fine fighter leader and was the ideal choice for the job. On his capable shoulders, and those of Johnny Walker, would rest the responsibility for leading most of the patrols in the battles to come. The striplings of Zurich would soon be coming into their own.

For the next three months the weather was appalling; in January the Squadron flew a total of 58 hours on the only five days when flying was possible. The aerodrome was unusable for many weeks, and a number of tailwheels were broken by taxying on the frozen rutted ground. 'Boy' Mould was detached to Etain with a section of three aircraft to do some patrolling with No 73 Squadron; while there, he intercepted a Heinkel 111 but was robbed of an easy kill by faulty ammunition.

The high winds of the final week of February continued to keep the Squadron on the ground, but served to dry out the little grass field at Vassincourt. No 1 now began to feel that life would start getting exciting as the weather improved, and it was time to take stock of events to date. The balance seemed pretty favourable. Of the five enemy aircraft attacked, four had been destroyed and the fifth damaged. Against this, one Hurricane had been damaged by enemy action and three had been written off in accidents. The half dozen or so new pilots had been given experience of operational patrolling. Confidence had steadily increased with the knowledge that the German aircraft so far encountered were anything but invincible.

Sqn Ldr Halahan's constant emphasis on professional flying, his insistence on maintaining a flying training programme whenever patrolling and the weather permitted, his inventive brain and his refusal to accept anything but the best treatment for his Squadron had produced an efficient and close-knit unit. Nos 1 and 73 knew they would soon have to face, virtually unaided, a force of over 3,000 German aircraft. At least in courage and skill, in determination and sheer guts, they would not be found wanting.

CHAPTER EIGHT

The Pace Quickens

March arrived in a great gale of wind. The last puddles dried from the aerodrome, and the morning of 2 March brought blue skies and a full day of patrolling on the right front between Nancy and Metz. At 10.45hr the patrol line was taken up by Hilly Brown's section with Sgt E. J. Soper flying No 2 to Brown as usual. Plt Off Mitchell, one of the new boys, was in the No 3 position. After an abortive distant sighting at Pont-à-Mousson they were turning over Metz when Soper saw a Dornier flying west at 24,000ft. They pulled their boost override plugs and climbed, flat out, to get above and behind the target. Soper got there first and engaged the Dornier, which made a hard 180° turn and dived steeply for the ground with the Sergeant's Hurricane in hot pursuit. Brown was about to attack when his wooden propeller disintegrated. He had to glide the 30 miles to Nancy airfield for a dead-stick landing, leaving his compatriots to finish off the German. Soper was meanwhile attacking the Dornier from astern, when Mitchell came roaring in on the beam and collected a burst from the rear upper gun, which knocked out his engine. As Mitchell broke away pouring white smoke, Soper pumped his remaining ammunition into the burning Dornier and turned back to give help. The smoke was obscuring Mitchell's view so Soper picked him a likely field and tried to talk him down, but the crippled aircraft's radio was switched to transmit, and Mitchell could not hear his advice. He crashed on landing and was killed.

That afternoon the Squadron patrolled the frontier from Thionville to Boulay with No 73, but nothing was seen. The news that Soper's Dornier had crashed did little to brighten the Mess at Vassincourt that night, and only the safe return of Hilly Brown did anything to cheer them up.

Brown and Soper were off on patrol again at 11.20 next morning and Joe Soper's keen eyes picked up a Heinkel 111 well above them at about 26,000ft. Once again they climbed at full throttle with boost plugs pulled. This was probably the highest combat that had yet been fought, and they had great difficulty in bringing their guns to bear. The German had realised their problem, and was climbing for all he was worth when Soper eventually opened fire at 200yd. Big chunks flew off the Heinkel and it dived for the comparative safety of the ground. Both engines were smoking from the continuing attacks, and the pilot eventually decided that he'd had enough, crash-landing in front of the Maginot Line in no-man's land.

Until now, virtually nothing had been seen of the German fighters; the Squadron was itching to cut its teeth on some real opposition. The first opportunity came on 29 March, when 'Pussy' Palmer, Paul Richey and Peter Matthews went to investigate some French ack-ack bursts which they noticed when patrolling over Nancy and Metz. They spotted the target, but had just lost it in a cloud at 25,000ft when Richey saw two sleek shapes pass close overhead in the opposite direction — Bf109s! The section whipped round and were climbing to the attack when Matthews saw three more closing in on Palmer from astern. He called him to break hard to starboard. Palmer did so, too hard, and spun down out of control to 10,000ft. Richey, meanwhile, had singled out one of the original Bf109s and was manoeuvring to position

The crewroom at Vassincourt — Prosser Hanks, flanked by Clisby and Lorimer, studies an important intelligence report. *Squadron Archives*

himself for the kill. Eventually the Bf109 pulled into a steep climb. Richey cut the corner and nailed it with a good burst. It fell out of control, smoking badly, while Richey managed to evade the others and returned safely to Vassincourt. He was credited only with a probable kill, as his victim fell on German soil without witness. German records show that Richey's Bf109 came from the crack *III/ Jagdgeschwader 53*, commanded by Werner Molders. The pilot was wounded.

The news of No 1 Squadron's first fighter combat since 1918 delighted everyone, but they were even more intrigued to hear that the C-in-C had offered a dinner in Paris for the first pilot to destroy a Bf110. This was the Luftwaffe's vaunted twin-engined fighter with which No 73 had grappled inconclusively a few days previously. All pilots now resolved to keep their eyes peeled with added zeal in the hope of winning this prize.

After lunch, Johnny Walker took off with Fg Off Bill Stratton and Sgt Arthur Clowes, with orders to patrol Metz at 25,000ft. As they reached patrol altitude, Clowes called his leader's attention to a group of aircraft approaching the section from the east and slightly below. As they were still some distance away, Walker turned the section towards them and strained his eyes to identify them. There they were — nine Bf110s, flying in loose sections of three. The recognition was mutual and the Messerschmitts broke into a hard climbing turn to engage the Hurricanes. Walker, too, turned as hard as he could and managed to swing his section into the rear of the enemy formation, but both Clowes and Stratton blacked out in the turn and lost several thousand feet in the process. Clowes recovered at 18,000ft with two Messerschmitts on his tail, leaving Johnny Walker all alone to cope with the other seven. After some hectic turns Walker singled out a Bf110 and was shooting bits off it when Stratton reappeared and also had a go, before it vanished into cloud. Clowes, meanwhile, was having fun. He managed to shake off his two pursuers, turned inside them and fired three short bursts at the leader, who stalled and went into a spiral dive. Avoiding the second Bf110 he dived for a third but could not catch it. Then, seeing Walker and Stratton busy with their victim, he fired on a

No. 2030, MAY 22, 1940] THE TATLER

"MEL" WITH THE A.A.S.F.—No. 1

NO. BLANK FIGHTER SQUADRON, R.A.F., SOMEWHERE IN FRANCE: By "MEL"

All these young pilots are fine-looking fellows and as keen as mustard, and they all have, to quote their C.O., "seen the black cross": in other words, they have engaged the enemy in aerial combat. The R.A.F. nowadays talk about their daily "ten before breakfast", and if we may judge by recent happenings, this is no idle boast. The Air Force has shown that it is as good as ever! This cartoon is the first of a series which our artist has been privileged to do in the actual zone of conflict.

A cartoon of some of the squadron's personalities, drawn by 'Mel' of the *Tatler*. *Squadron Archives* fourth before overshooting the combat and escaping into a cloud. The section, now out of ammunition, very wisely beat it. They joined up at Metz before returning to base, where they heard that Walker's Messerschmitt had been found in pieces near Bouzanville. The section's combat report caused great excitement and rejoicing: they had emerged unscathed and victorious from a dogfight with Germany's supposedly top air superiority fighter at odds of three to one against. In addition to the confirmed kill, Clowes thought that he might have damaged two of his adversaries.

At the intense debriefing session that night they all agreed that the Bf110 was faster than the Hurricane in a dive but was not so manoeuvrable. They also reckoned that the rear gunners had difficulty in aiming in a tight turn, as the Hurricanes had not come under fire when attacking from astern. Clowes had also noticed what seemed to be heavy calibre tracer fire coming from the propeller shafts of the engines, which could possibly indicate additional cannon armament. All these details were passed to RAF Intelligence and were circulated throughout the RAF, helping to relieve some of the apprehension about the Bf110.

There now arose a delicate point of diplomacy. The C-in-C had offered dinner to the one pilot who destroyed the first Bf110; but Johnny Walker insisted that it was to be three or none. Bull Halahan took a deep breath and sent the following signal to AM Barrett: 'Regret that dinner will have to be for three, as pilots say Bf110 brought down by teamwork. Please inform'. The reply came back: 'Dinner for three it is. Will send aircraft'. The Air Marshal's personal aircraft, a Percival Q6, arrived to collect the three victors on the 31st. They flew to Paris and dined with the C-in-C at Maxim's — where else! The Squadron record book relates that a corking good time was had by all.

While the well-fed trio were sleeping off their excesses on 1 April, Prosser Hanks was leading Boy Mould and Digger Clisby on a left front patrol up near the corner of Luxembourg. They spotted another formation of nine Bf110s 4,000ft above them. Despite the unfavourable position and odds, Hanks decided to attack.

'As we were climbing below and a little to one side of the enemy, they saw us and turned to the attack. Three of their aircraft remained in position, while the rest of them followed the leader. The leader turned straight for me, and he and I fired at each other head on. His burst was cannon and incendiary bullets, and one cannon shot penetrated my port leading edge, went through the main spar, the oil tank and petrol tank, and shrapnel pierced the perspex hood. Oil poured out of the main plane, but I decided to finish my ammunition off before leaving the combat. I fired at about four different aircraft, and saw one go spinning down completely out of control, but as I watched it go down, another got on my tail so I lost sight of it.'

The tenacious Hanks managed to evade this last attacker, and limped his crippled Hurricane to a safe landing at Etain. Mould and Clisby were still very much in the fray, giving each other continuous cross-cover while they sent two more Messerschmitts to earth in flames. They finally had to fight their way clear of the three Germans who had remained above the battle, before landing at Etain completely out of ammunition. All three victories were later confirmed by wreckage on the ground.

As Johnny Walker was away for the day, Clisby led both sections of 'A' Flight to patrol Metz on 2 April. They had sighted a Dornier and were in hot pursuit when Pussy Palmer called up to say that three Bf109s were diving on the Hurricanes from behind. Clisby turned the leading section in time to see Palmer being attacked by the first Bf109. The German pilot fired at point blank range and set Palmer's reserve tank on fire. As the flames started entering his cockpit, 'Pussy' rolled on to his back and baled out, free-falling several thousand feet to avoid the rest of the dogfight before opening his parachute. He finish up in a tree in no-man's-land, some 300yd from the foward German positions, but was fortunately rescued by a French patrol from the famous 'Diables Rouges' regiment. The soldiers entertained him royally before driving him back to Vassincourt.

Clisby, meanwhile, had latched on to the tail of the victorious Bf109 and shot it out of the sky, following the blazing wreckage down until it vanished into cloud. Kilmartin, who had been flying No 2 in Palmer's Yellow Section, shot down another 109 which had foolishly flown right across his nose. Killy also followed his victim down, but was saddened to see it plunge into the middle of a French village.

Throughout the next week, continuous cloud cover prevented further German reconnaissance. It was against the rules to send pilots across the border to look for trouble so the flying effort was devoted to the training of the new pilots. Salmon and Shaw, who had arrived in March, were joined by Plt Off Peter Boot. The three of them were sent off on the wingtips of a succession of the more senior pilots to learn the tricks of the trade. Lewis and Matthews were now shaping up well, and had flown on a number of patrols. They were sent off together for a week to practice attacks and escort techniques with the Battle squadrons at Mourmelon.

On 11 April, the Squadron stood by to greet Marhsal of the Royal Air Force the Viscount Trenchard, who was touring units in France and was expected to arrive at Vassincourt at 11.00. The great old man never came, for the Squadron was suddenly ordered to move immediately to Berry-au-Bac: Hitler had invaded Norway. The AOC was afraid that the invasion of France would soon follow, and needed close cover for the light bomber squadrons based at airfields around Reims. By 17.45 the Hurricanes had refuelled and were standing by at their new airfield alongside the Fairey Battles of Nos 12 and 142 Squadrons. The ground party and MT arrived throughout the night, minus the Leyland coach, which had crashed en route, fortunately without injuring its occupants. By morning the troops were billeted in a couple of nearby villages and the Battle pilots had made room for the officers in their Mess at the Chateau de Guignicourt. These domestic arrangements were far from ideal, for elements of the Squadron were separated from the airfield and each other by several miles without so much as a telephone between them. It was obviously time for the CO and his faithful interpreter to swing into action again! By blackmail or bribery, Sqn Ldr Halahan had managed by the evening of the 16th to have the entire Squadron moved into the village of Pontavert. As well as being very close to Berry, Pontavert had been built since 1918 and was clean, bright and friendly.

Continuing cloud and a quiet front soon removed the feeling of urgency that had been created by the move. Little flying was done for some days, and the only item of interest was the arrival from the factory of a Hurricane fitted with the new constant-speed airscrew, which was immediately flown by the CO and a number of the more experienced pilots. The new prop proved to be a remarkable improvement over the previous wooden-bladed variety and gave the Hurricane a ceiling of 30,000ft from where the Bf109s could now be engaged in equal terms.

A dispatch rider had been sent back to pay the unit's bills at Neuville and Bar-le-Duc, and they had just begun to settle in at Berry-au-Bac when orders came through for No 1 to return to Vassincourt with maximum secrecy on the 19th. The vehicles were sent back individually during the morning and were followed in the afternoon by the Hurricanes, which flew alone and by devious routes. The aircraft were turned round and on state at Vassincourt by 17.15. Everyone was pleased to be back, although the officers had some difficulty in ousting the Mayor from their Mess — he had repossessed it within hours of their previous departure. That evening the pilots were guests of the Ensminger family at an excellent dinner given at the Hotel de Metz.

At 09.30 the following morning a group of bleary-eyed pilots of 'A' Flight were lounging about after an early patrol when a Junkers 88

Plt Off Billy Drake, who chased a Bf109 under some power cables before shooting it down on 20 April 1940. *IWM*

was seen directly overhead at about 20,000ft. Realising that it was probably photographing the airfield, Kilmartin and Richey hurriedly scrambled in pursuit. Paul Richey, flying one of the old wooden-bladers, had his propeller disintegrate without warning and had to glide home. Killy, after being fired on by some French ack-ack, eventually caught up with the Junkers near Reims. The crew saw him coming from miles away and continued westward, taking evasive action in a series of great swoops and climbs. This enabled the Hurricane to catch up even faster, but by the time that Killy had closed to firing range both aircraft had descended to ground level. He let fly with a withering burst of fire that put the German's starboard engine out of action, but the rear gunner was also on his toes and put a bullet through Killy's cylinder block, causing him to make an immediate belly landing on the outskirts of Paris. The Junkers 88 subsequently crashed many miles to the south, near Macon. It had been the crew's first foray into France and they had been irretrievably lost. 'Killy' was rescued unhurt by the French Army and was returned later in good spirits.

Later that morning, Johnny Walker led four aircraft of 'A' Flight on a 'left front' patrol and spent some time chasing a small aircraft, which turned out to be a Spitfire, probably from No 212 Squadron at Tigeaux. Soon afterwards they ran across nine Bf109s straggling along in a haphazard fashion. The Germans were taken completely by surprise. Walker dived in and clobbered the leader with two short bursts and Hilly Brown claimed one of three which had tried to escape into Germany. Billy Drake, meanwhile, was having a tremendous duel with three more Bf109s. One of them broke away, dived to ground level and headed back across the frontier. Drake gave chase.

'One evasive tactic which impressed me not a little was to fly very low until he saw some high tension wires, under which he flew. I saw them in time, however, and also managed to fly under them. I then found myself flying out of range (of the Me 109) in a valley, when he suddenly turned to the left and gave me time for a perfect deflection shot. Black smoke poured forth, and he dived straight into a hill and burst into flames.

'During all this the third German was endeavouring to get on my tail, but fortunately failed. I returned to the aerodrome via the Siegfried Line, and was shot at by pom-poms, which also missed.'

The rest of the month was fairly uneventful, and the CO took the opportunity to hold a series of discussions on tactics. It had now become obvious that the old 'air drill' concept was quite useless in combat because it left the tails of the formation completely unprotected against a bomber's escort. Over the previous weeks a number of different attacks had been tried, and they reached the conclusion that the ideal formation consisted of five Hurricanes with a spearhead of three, covered by two 'arse-end Charlies' whose job was to weave over their tails and to warn of any attack from the rear.

They only met the enemy once more before the end of April, when Prosser Hanks was leading a patrol of five 'B' Flight aircraft on the 22nd. Near the frontier they met 15 Bf109s — almost two full *Staffeln* — flying well above the Hurricane's absolute ceiling. After some circling and jockeying for position, two of the Messerschmitts were sent down to attack, but were driven off into Germany with ease by Mould and Clisby. The remaining Germans, although still considerably outnumbering the Hurricanes, did not come down and fight, eventually following their friends across the frontier. It was assumed that Hanks and Co had run into a group of trainee pilots cutting their combat teeth under supervision, but the encounter indicated some of the respect that the Luftwaffe was beginning to show for the tiny force of British fighters that opposed them.

By the end of April all the Hurricanes had been equipped with constant speed propellers, but the new aircraft had to be collected from a Maintenance Unit at Amiens. A number of pilots were ferried across by Moses Demozay in the Miles Magister that the Squadron used for communications purposes. One one occasion he was taking Billy Drake to the MU to collect a new aircraft; his passenger was very tired and slept all the way. On arrival Drake noticed that the Magister had been damaged en route, and asked Moses what had happened. The Frenchman explained that they had been attacked by Bf109s, and he had flown through a tree while taking evasive action!

On 3 May it was learned that a Bf109 had been captured intact. The CO took a flight of six aircraft over to inspect the prize at Amiens, and Hilly Brown was invited to fly it. He and Prosser Hanks later fought a low level combat over the airfield to evaluate its performance, and several valuable facts emerged. The Bf109 had excellent rearward vision and proved to be unquestionably faster than the Hurricane at

Boy Mould, Bull Halahan and Leslie Clisby with some of the trophies in the Officers' Mess at Neuville shortly before the German invasion of France.
Squadron Archives

high altitude. To offset this, the Hawker machine was more manoeuvrable and slightly faster on the deck. After the evaluation, Hilly flew the Messerschmitt to Boscombe Down with a heavy escort. He returned to Vassincourt the next day just in time to put his knowledge of the 109 to good use on 7 May, when he shot one down in conjunction with Plt Off Roland Dibnah.

Dibnah, together with 'Red' Stavart, Don Thom and Randy Goodman, had recently arrived from 6 OTU at Sutton Bridge. Halahan was not prepared to integrate them into his fighting team at this stage, as all the signs were pointing to an imminent German invasion. There would be plenty for them to do on the ground in the two weeks that followed.

And still the Luftwaffe did not come. Far to the north, vapour trails could be seen daily by the pilots on patrol, but the skies over France were uncannily quiet. At Vassincourt the promise of a fine summer wafted the scents of the countryside over the crouched shapes of the waiting Hurricanes. The air was full of wild rumour, and a feeling of unease began to spread that this must be the lull before a mighty storm.

Then, late on the evening of 9 May, they heard the low rumbling of heavy artillery from across the northern horizon: the great guns of the Maginot and Siegfried Lines had begun to speak. The pilots bade each other goodnight, and walked silently to their billets. It had come.

CHAPTER NINE

Ten Days in May

10 May 1940

'B' Flight was called at 02.15hr to be at readiness when dawn broke soon after three. It was scrambled at 05.00 and 'A' Flight was dragged out of bed to replace it on standby, because the observers in the Maginot Line were reporting large formations of enemy aircraft passing overhead.

Flt Lt Prosser Hanks. Promoted to lead 'B' Flight at the age of 22, Hanks with Johnny Walker must take the credit for the survival and success of No 1 as a fighting team during the withdrawal through France, and he destroyed at least nine German aircraft himself. *IWM*

'B' Flight, led by Prosser Hanks, sighted a formation of Dornier 17s and swept into attack the final section of three. Hanks shot down the leader, while the right hand Dornier was damaged by Mould, who overshot his target and got badly peppered for his pains. The right leg of his flying suit was ripped by a bullet and one of his tyres was punctured, but he was able to land safely.

'A' Flight was scrambled soon after 'B' to patrol Metz at 20,000ft, and destroyed a Dornier 17z. The pilots landed at Etain to refuel and learned that their victim had been bombing the railway station at Longuyon. Walker had been hit in the engine during this engagement and landed wheels-up near Verdun.

During the morning, orders were received to move the Squadron to Berry-au-Bac to look after the light bombers once again. 'B' Flight was on readiness at the time and, if scrambled, was to engage and land at Berry, while 'A' Flight helped to pack everything up at Vassincourt.

As they awaited the order to scramble, the pilots could hear heavy bombing in the distance, and enemy formations passed tantalisingly close overhead. But their orders were clear — the primary task was bomber escort, and passing Germans were not to be attacked indiscriminately. In the event of a raid on Vassincourt they were to take off and return when all was clear, leaving the French fighters to cope with the Luftwaffe.

At noon 'B' Flight was ordered to fly to Berry-au-Bac. On the way it ran into a mixed formation of German bombers and attacked them. Hanks shared a Dornier with Lewis, while Mould and Drake each brought down a

Heinkel 111. They landed at Berry-au-Bac minus Fg Off Lorimer, who had been hit and had baled out.

'A' flight was led to Berry by the CO after lunch, and the Squadron spent the afternoon settling in at its new aerodrome. Soon after arrival, it was bombed by Heinkels from high level. A stick of 14 anti-personnel bombs landed too close to the dispersal for comfort and three French labourers working in a nearby field were killed. Their horses were badly injured and had to be destroyed by the CO and Leslie Clisby. The remaining pilots were more upset by this incident than by anything else that had happened since they had arrived in France.

Earlier in the day Fg Off Salmon had been sent off in his Lagonda to collect Walker and Lorimer. He returned them safely to Pontavert in time for supper.

11 May 1940

The pilots were woken at 02.45hr to prepare to prevent the Luftwaffe knocking out Berry-au-Bac. A heavy raid took place on Reims, but the airfield was not attacked. At 05.00 they were stood down in sections for breakfast and later spent some time rigging up telephone lines, tents and chairs at the operations tent in the woods.

The early raid on Reims had been directed against the headquarters of the AASF, and the staff officers who had led a life of peaceful luxury for the past six months had to take shelter in the champagne cellars of their chateau. Their discomfort resulted in a slight change of strategy, for Halahan was now ordered to be ready to lead the whole Squadron against any future raid that threatened the bases or headquarters of the AASF.

As he had previously been forbidden to fly operationally by the AOC, the Bull was quite amused by his order. To his relief, it also meant that he would be able to take the place of one of his Flight Commanders if the need arose.

At 08.00hr a dozen Heinkels were reported to be approaching Berry-au-Bac, and the Squadron was ordered into the air. The CO told 'A' Flight to circle the airfield in defence while he led 'B' Flight up to intercept the raid. The bombers saw them coming, turned and climbed away into the sun at full speed. After a few minutes of fruitless pursuit, Halahan turned his aircraft back to the airfield. Paul Richey meanwhile was chasing a Dornier 17z which he had spotted. He damaged both its engines badly, but then became lost. He landed at Mezieres, where he hit a bomb crater and damaged his aircraft.

At 16.00hr 'B' Flight took off to escort a Battle squadron to the northeast of Rethel, and ran into a flock of Bf110s at 17,000ft on the way to the target. Hanks, Mould and one other pilot each claimed a Messerschmitt; Clisby landed on a French airfield and was shot at by the defences in the process.

Soon after they landed, a heavy raid was reported to be approaching Reims, and 'A' Flight was scrambled to intercept at 10,000ft. The five Hurricanes were over Mezieres when they saw 30 Dorniers covered by 15 Bf110s. Walker led his aircraft above the enemy, they dived on the escort from astern, and a tremendous dogfight resulted, in which Brown, Kilmartin and Soper each destroyed two Bf110s. Walker claimed one destroyed and one probable. Richey took on five of them single-handed and shot down two, before being hit in the front fuel tank by a cannon shell. He baled out with considerable difficulty, and was rescued by the French Army. Walker's probable must have crashed, as 10 wrecks were later found on the ground, and the enthralled witnesses below had seen 10 Messerschmitts fall in flames.

When questioning the pilots of 'A' Flight after this battle, it was extremely difficult to get a clear view of what exactly had happened: after dog fighting themselves into a stupor, they were still pressing imaginary buttons and pulling plugs an hour or so after landing. It should be added that the Dorniers, deprived of their escort, had immediately headed back into Germany.

12 May 1940

After the dawn readiness, the pilots of No 1 Squadron were held in a state of frustrated idleness for several hours. The Luftwaffe continued to pound at railway stations and other airfields near Reims, but only two sections were scrambled to intercept them during the morning.

The rear party of 100 men, left behind at Vassincourt in case the Squadron returned, was now ordered to set out for Berry without delay. The Germans were advancing at an amazing speed through Belgium and the Ardennes, and no longer was it thought likely that the Squadron would be returning to Vassincourt. In Belgium, the advancing Wehrmacht had captured two vital bridges across the Maas at Maastricht. The retreating Allies had failed to blow them up, and they were now carrying the bulk of two Panzer divisions across the river.

Late on the previous evening, Sqn Ldr Halahan had been visited by Flt Lt 'Judy' Garland of No 12 Squadron. Garland said that he had been tasked to bomb the bridges at 09.00hr with six Fairey Battles. Volunteers had been asked for in view of the extremely dangerous nature of this mission and the whole Squadron had stepped forward.

Garland considered that his task had been very badly timed because the morning sun would obscure the German fighters attacking from the east, and would give the flak gunners a good view of the Battles as they approached from the southwest. He had planned to approach the target at very low level, but would have to pull up to several thousand feet before diving on the bridges. He told Halahan that the Battle crews would feel a lot happier if they had a fighter escort over the target, and asked him if No 1 Squadron could help.

Later that night the request had been cleared by HQ AASF, and Sqn Ldr Halahan decided to lead the escorting Hurricanes himself. After some thought he decided to take a representative cross section of the Squadron's pilots with him to Maastricht. Hilly Brown and Leslie Clisby, his usual wingmen, would accompany him in the leading section; the remaining five aircraft would be led by Sgt Soper with Kilmartin, Lorimer, Boot and Lewis.

There were other, and more personal, reasons why the Bull wanted to lead this show. No 12 Squadron had been formed at Netheravon in 1914 by a detached flight of No 1: its commander had been Sqn Ldr Halahan's father.

At last it was time to brief. The CO planned to lead the Hurricanes north to Namur, from where he would follow the Allied-held western bank of the Maas to the target area. To cope with any enemy fighters he would try to reach the bridge at 08.57hr, three minutes before the Battles were due.

They met the first Messerschmitts over Namur. Several pairs and fours of Bf109s were flying about, but they seemed reluctant to press home their attacks on the tight little formation of Hurricanes.

Many more small groups of enemy fighters were encountered in the target area. Kilmartin recalled that they appeared to lack any united purpose and that the Squadron had almost reached the bridge before any fighting took place.

Halahan arrived over the bridge at 08.59 just one minute before the bombers were due. By now, some of the Bf109s were beginning to make firing passes at the Hurricanes, so the Bull made a wide sweep of the target area in the hope that the Battles would escape their notice: he had just spotted Garland and his men pulling up for their dive on the target, and could only hope that the gunners below had their gaze diverted by the twisting mass of fighters overhead.

It was not to be. As he drew the Messerschmitts away from the bridge, Halahan caught a glimpse of what happened. The Battle crews dived their slow and cumbersome bombers against the muzzles of 300 flak guns, and they were hacked out of the sky one by one. The CO saw Garland's aircraft blown to bits over the bridge. Its bomb had not been exploded by the flak, and fell on the parapet. Only three men of the eighteen involved returned to base, and 'Judy' Garland was not among them. He was awarded a posthumous Victoria Cross.

By now the action overhead was warming up. In the first hectic seconds of the dogfight, the CO, Clisby and Soper each brought down a Bf109, while Lewis, taking part in his first major action, destroyed two in rapid succession.

As the German attacks were now coming in from all directions, Halahan split the formation for individual fighting. Lewis's aircraft was set on fire so he baled out, minus one eyebrow. Two Bf109s were turning in his direction as his parachute opened, and Hilly Brown, fearing that they might take a crack at Lewis, dived in and opened fire on the leader. The two Messerschmitts collided and spun down, locked together.

The CO was hit in the engine by a cannon shell. The Merlin hauled him clear of the dogfight with its final gasp, and he started to glide towards the Belgian lines. On the way, he met two Arado communications aircraft head on. He shot them both down before crashlanding safely in Allied-held territory.

Peter Boot shot down a Bf109 soon after the CO had been hit, Clisby brought down his second Messerschmitt, and Killy destroyed a Heinkel 112. His identification of this rare gull-winged German fighter is beyond doubt, and is the only definite record of this type being used in combat service by the Luftwaffe.

By this time both Brown and Soper had left the fight with battle damage. Clisby, having the time of his life, destroyed his third Bf109 just before Lorimer got his first.

Killy was getting short of fuel, and the only Hurricane he could see was Leslie Clisby's,

Fg Off John Ignatius Kilmartin — Irish, and 'a Holy terror'. He shot down at least eight enemy fighters during the first week of the Blitzkrieg in France. *IWM*

which had dropped down behind a row of Arados that were bumbling along at low level. Killy heard Clisby whooping with delight over the R/T as he closed in behind and shot down the last three in the line. They then joined up and flew back up the Valley of the Maas to Berry, only Hilly Brown getting back to Berry-au-Bac before them. The others had run out of fuel and landed elsewhere for more.

The Bull had a number of adventures on his way back to base. He had forced-landed near a main road, and observed that some of the Belgian troops were running faster than the refugees. He joined the straggling column, but soon afterwards a Junkers 88 in French markings swept in and began to machine-gun the road. Halahan jumped into a ditch, escaping without injury; many of the refugees were not so lucky. Some time later, a Heinkel was shot down at the roadside. At the time a regiment of French Moroccan troops was passing, going up to the front on the trot, so a detachment of the Africans was sent over to the Heinkel. Halahan was later told by the French that they had decapitated the crew.

When he eventually arrived at Berry-au-Bac on the following evening, Halahan called his pilots together and told them of what he had seen. They listened with growing anger to his story of how the Germans were strafing the refugees to prevent the escape of the Allied armies. It all seemed so different from the Phony War of the previous months.

13 May 1940

'B' Flight was scrambled shortly after dawn to intercept a raid thought to be approaching Reims, but failed to make contact. While at altitude Billy Drake suffered oxygen failure and returned to Berry-au-Bac. On the way he spotted four Dorniers so attacked them, and had destroyed one, and probably another, when he was pounced on from behind by an escorting Bf110. It wounded him in the legs and back and shot his Hurricane to pieces. He baled out and was taken to hospital at Rethel.

On a later sortie Prosser Hanks shot down a Heinkel 111 which landed in the same field as one of the Bf110s that had been destroyed by the Squadron on the 11th. Otherwise it was a very quiet day. The sound of distant bombing was never absent at Berry, but little came their way and there were no escort missions to be flown.

'Leak' Crusoe and Sgts Berry, Albonico and Clowes, who had been recalled from leave in England, arrived back at the airfield in the afternoon. They were soon followed by the CO, as previously related, but of Billy Drake and Lewis there was still no sign. Late in the afternoon a Dornier in French markings swept low over the squadron's dispersal as the tireless groundcrew were working to repair the battle damage of the previous few days. They felt that this forecast the possibility of big trouble for Berry-au-Bac on the following day. After supper the news arrived that Lewis had been captured alive by the Germans, and then liberated by a Belgian counter-attack: Richey, still unfit to fly after his bale-out on the 11th, was sent off with a driver to pick him up from Maubeuge. The remaining pilots, dog-tired but suddenly happier, went off to bed.

14 May 1940

This time it was 'A' Flight's turn for the early standby, but no German aircraft came close enough to warrant a scramble. When Johnny Walker took his men to breakfast, 'B' Flight strapped in and waited. At 06.00hr they were scrambled, and flew headlong into a large formation of He111s escorted by at least two squadrons of Bf110s. The escort was engaged first and three were quickly shot down by Mould, Clisby and Hanks. Astonishingly, the surviving escorts turned tail and fled, but not before Hanks had added another to his growing score. The five Hurricanes now had a free hand to deal with the bombers. Hanks led Lorimer and Goodman in on a classic stern attack which sent one down in flames. Boy Mould also brought down a Heinkel, while Clisby took on two of them and forced one to land in a large field. Seeing the crew climbing out of the wreck, the Australian landed alongside and gave chase, waving his revolver! The Germans, fearing for their lives, surrendered immediately, and were handed over to the French Army. Clisby then took off and returned to Berry.

For the previous two days the German Army had been pounding on the gates of France at Sedan, supported by hundreds of Junkers 87 dive-bombers. There, as at Maastricht, they had to cross the river, and the Battles were again ordered out to try to slow the German advance. 'A' Flight was assigned to escort the Battles this time, but they ran into a flock of Stukas escorted by Bf109s. Again the escort was attacked first, and again it fled before the determined assault of a handful of Hurricanes, leaving behind six of its number smoking on the ground. With whoops of glee 'A' Flight fell upon the unprotected dive-bombers, but only had time to shoot down three before breaking off to escort the Battles back to base. Fortunately for the Stukas this was the only occasion they met No 1 Squadron in the early days of the Blitzkrieg.

During all this excitement, Fg Off Salmon had been sent off to Mezieres with a party of men to repair the machine which Richey had landed there on the 11th. They arrived in the middle of a heavy air raid in time to see the Hurricane destroyed in a strafing attack. During the raid there was no sign of any French pilot, and the airfield staff had scattered wildly into the surrounding countryside. This confirmed another of the Squadron's suspicions — that many units of the French Air Force were not doing their job. They were supposed to be intercepting the bulk of the incoming raids, yet not one French fighter had been seen in the air since the start of the Blitzkrieg. Every Curtis and Morane fighter at Mezieres was found to be immobilised under camouflage netting.

By now the stream of refugees passing the gates of Berry-au-Bac had grown to a flood. The news of Rommel's armoured break-through at Sedan had spread and the French were leaving their homes in tens of thousands. This made it virtually impossible to resupply the front-line units and airfields. At Berry there was fuel and ammunition only for a further six days fighting, but by then, who knew whether the Squadron would still exist? So far the survival of its pilots had seemed nothing short of miraculous and this was matched only by the incredible list of victories that was piling up day by day. But for how long could this go on? With every sortie flown their fatigue was biting deeper. Only time would tell.

15 May 1940

Berry-au-Bac was raided lightly several more times during the morning, but the Hurricanes, well dispersed and camouflaged in the wood, escaped damage. 'B' Flight, having taken the punishment of the dawn raids, was relieved at 08.00 by 'A', which remained on readiness until 11.30hrs. It was then scrambled by Wing to meet a heavy incoming raid at Vouziers. Paul Richey was now fully recovered and flying again for the first time since his bale-out on the 11th. In his book he wrote:

'Five minutes later Johnny said: "Christ! Close in — Close in!" and we saw them. There were about forty Dorniers in close formation at 10,000ft, and above them squadron after squadron of Me110s stepped right up to 18,000ft. I counted eighty and then gave it up, for everywhere I looked I saw another squadron of them. There must have been well over a hundred and fifty Huns in all. The whole lot were coming straight at us.

'Johnny led us to the left, climbing as hard as we could. As we skirted around their right flank, Johnny turned towards one of the Me110 squadrons at 16,000ft, saying "Here we go — and keep a bloody good look out!" Fortunately, I was doing so, and saw another Me110 squadron 2,000ft higher behind us as we were going in. I shouted over the R/T to indicate them. Johnny answered, "OK Red 3! OK!", and we swerved away and climbed up again flat out'.

As they turned in behind the top squadron of Messerschmitts the six Hurricanes were spotted. The enemy immediately formed a defensive circle, and 'A' Flight broke up to make individual attacks. Richey and Walker each shot down one Messerschmitt before the circle was broken, but by that time another squadron had been alerted and was climbing to the rescue of its comrades. Kilmartin had just sent one down in flames when he was attacked from behind. He was rescued by Richey, who shot down his assailant before being badly hit himself. Richey dived headlong out of the dogfight with his glycol tank on fire and had to bale out again before getting back to Berry.

Meanwhile the five surviving Hurricanes were fighting for their lives. As the twisting mass of aircraft descended from 18,000ft, more and more Messerschmitts were able to join in. Walker had managed to shoot down another Bf110 before stopping a bullet in his engine, and had to dive away to force land in a field. Soper, Clowes and Hilly Brown each claimed one of the Germans before they managed to break free and return to Berry-au-Bac.

After lunch Sammy Salmon was sent off on another of his errands — this time to drive three new pilots over to No 501 Squadron at a nearby airfield — and witnessed a fight between some Bf110s and six Hurricances of No 73 Squadron. He said later that if he'd had a ciné camera he could have made 'Hells Angels' look silly as the whole scrap was conducted in and out of the treetops. He captured a German pilot who had baled out and was able to take Fg Off 'Fanny' Orton to hospital. Orton had jumped from 600ft wuth a cannon shell in his shoulder.

Soon after tea, six aircraft of 'B' Flight were sitting at readiness when 30 Bf110s flew overhead. They had not been ordered to scramble but one of the fitters begged Prosser Hanks to go up and get them, saying: 'No 1 Squadron's not going to let the cheeky sods get away with that! Come on Sir! I'll start all the bleeding kites up myself if you'll go!' With that, they scrambled and chased off towards Laon in pursuit.

Over Laon the Messerschmitts turned, spotted them and swung back into the attack. Very soon there were six individual battles going on, at odds of anything up to 10 to one. Boy Mould managed to shoot down two before running out of ammunition and legging it back to base. Hanks dived on the leader, and blew him to bits with one burst, but was then set on by several others. One of them punctured his glycol tank and almost blinded him with the hot fluid: he baled out and landed in open country with eyes like poached eggs. Boot and Lewis had knocked down one each before running out of ammunition; of Clisby and Lorimer there was no sign.

By now the floodgates at Sedan were fully open and the tide of German armour was sweeping into northern France. Although the refugees were still streaming past the airfield, blocking every main road, the Squadron was visited in the evening by AM Sir Edgar Ludlow-Hewitt, who had led No 1 Squadron's very first operation in Flanders in 1915 with Eustace Grenfell. Now Inspector-General of the RAF, he listened in amazement and pride at the Squadron's achievements of the past few days. But it was dusk and there was no time to spare even for the most distinguished of visitors. Wearily, they sank on to their camp beds. Two of the beds were empty, for Clisby and Lorimer were never seen again.

16 May 1940

Just after midnight, the CO sent a runner to wake every member of the Squadron. He had been on the telephone since they went to bed, and the news was bad. The German advance was ploughing through France faster than anyone could believe. Rethel had fallen soon after Billy Drake had been evacuated from the hospital and if the AASF was not careful, its airfields would soon be cut off or overwhelmed by any German thrust to the south. At midnight the Bull had received his orders — No 1 Squadron was to move to Vraux without delay. The new airfield was at Condé-sur-

Fg Offs Lorimer and Clisby were last seen attacking a formation of 30 Bf110s on 15 May 1940. On the night before he died, Leslie Clisby tallied his kills with Sqn Ldr Halahan, and arrived at a figure of 19 aircraft — 14 of them within the previous three days. *IWM*

Marne, halfway between Reims and Paris: if Guderian's Panzer divisions should choose to swing down to take the capital, it would lie squarely in the path of the German advance.

Sqn Ldr Halahan had been grounded again by the C-in-C after the Maastricht raid. As a result, he had decided to lead the advance road party to Vaux, and set off at 01.30hrs. The Hurricanes flew across shortly after dawn, and the pilots spent the rest of the day at cockpit readiness, waiting for the German assault on Paris. It never came. Unknown to them, the enemy's objective was the Channel ports, and there was little activity to the south of its armies all day.

The rear ground party had left Berry-au-Bac just in time; French engineers blew up the bridge over the Marne-Aisne canal just after the Squadron's last lorry had crossed.

No 1 Squadron shared the field with the Blenheims of No 114 Squadron, which threw open its Mess to No 1's pilots. No 114 had taken a severe beating a few days previously, and only three of its aircraft remained intact. No 114 Squadron's Mess was a small estaminet pleasantly set by a canal near the airfield. After dusk the pilots arrived there for supper — they had been on patrol or at cockpit readiness all day and the fatigue was really beginning to bite. The CO recalled that they each took one glass of red wine with the meal, and the next thing he remembered was being shaken awake by the Adjutant some time later. Every pilot had fallen asleep where he sat.

Sqn Ldr Halahan woke his men and sent them off to bed. He then sat down and wrote a dispatch to the AOC. If this pace was allowed to continue, he wrote, his pilots would all be killed in action within a week because of their state of complete exhaustion. He requested that all the original men who came to France in September be repatriated without delay. Their skill and experience would be needed by Dowding in Fighter Command in the very near future: France was already lost.

The CO gave the letter to his dispatch rider and looked forward to a good night's sleep. Even this was to be denied him.

17 May 1940

During the night a column of French tanks charged through the village of Condé-sur-Marne with guns pointing firmly rearwards. They stopped just long enough for the troop leader to tell the CO that the Germans were already at Reims, and would arrive shortly before dawn. For the next hour the Bull tried desperately to get through to No 67 Wing or Headquarters AASF for instructions, but the telephone lines were down. He therefore had the Squadron roused at once, and ordered the advance party to set out without delay to a camouflaged airfield that he knew of at Anglure, some 30 miles away to the southwest. He followed on himself at 03.00hrs, taking the spare pilots along in the Humber and leaving the aircraft in the capable hands of Johnny Walker, who was to follow on at dawn.

No sooner had he left than No 67 Wing got through on the telephone, and told Walker that the French cavalry must have the wrong end of the stick. The Squadron, it said, was to remain at Vraux. With fuel and ammunition for one day's fighting only, a tiny handful of groundcrew and no spares, Walker decided to sit it out and await developments. As dawn was breaking he put 'B' Flight on standby and sent his own pilots back to bed for a few hours.

The Flights changed over at 09.00. Walker led five Hurricanes off on a sector patrol to find out where the Germans were. There was no sign of any enemy troops in the area, but they ran into a formation of Heinkels escorted by a dozen Bf110s. Soper brought down two of the escorts on fire, but got so badly shot about by one of the German rear gunners that his aircraft landed looking like a Swiss cheese. Killy also got a brace, and saw 'Pussy' Palmer shoot one down before baling out of his blazing Hurricane. Hilly Brown then destroyed the Bf110 which had got Palmer, and also managed to shoot down one of the Heinkels before returning to Vraux.

In the afternoon a section of 'B' Flight was scrambled to chase off a formation of five Dorniers which were circling overhead. Peter Boot got there first, and was amazed to see them split up and race away to the north on his approach! This tended to confirm the story that was going about at the time that 'Cobber' Kain of No 73 had turned in to attack a Dornier, but the crew had baled out before he could fire a shot. So much for the courage of the invincible Luftwaffe!

By late afternoon the AOC rang up and confirmed that No 1 was to move to Anglure after all. The German movements were still uncertain and the bulk of the groundcrew were on their way there without hope of recall. The knowledge that the Squadron was to be reunited the next day came as a great relief to Walker, because his supplies of fuel and ammunition were now perilously low. Everything was packed up and ready to go by nightfall, and this time the Squadron was allowed to get an undisturbed night's sleep.

18 May 1940
The 12 serviceable Hurricanes took off for Anglure at 06.00. The rear ground party pushed Soper's shredded aircraft into a bomb crater, and gave it a full military cremation: many of the Hurricanes which had been damaged in the past week had been abandoned in this way, or had been left in the countryside where they had landed with minor scars. Under normal circumstances most of these could have been repaired by station workshops within a week, but now there were no such things as workshops, and no such thing as a spare week.

Only 11 Hurricanes arrived at Anglure. Plt Off Stavart, one of the new boys, had seen a Dornier soon after taking off from Vraux and had shot it out of the sky. Delighted by his first kill, he set off at a rate of knots in the direction

Sgt Plts Berry and Albonico at Neuville before the invasion. Albonico was shot down and captured on 18 May, but Bill Berry remained with the Squadron throughout the Battles of France and Britain until his death in combat on 2 September 1940. He shot down the Heinkel that sank the troopship *Lancastrian* at St Nazaire. *IWM*

in which he fondly imagined he would find Anglure. At the appointed time an airfield appeared and he prepared to land, then deduced from the sight of a Heinkel landing ahead of him that the airfield was German. He promptly shot the Heinkel down and pushed off at high speed. He tried to make it back to Vraux, but his fuel ran out a few miles short. He landed in a ploughed field and arrived at Vraux a trifle incoherent, just as the groundcrew were leaving.

Anglure aerodrome consisted of a collection of fields outside the village of Pleurs. Its unprepared surface gave it an excellent natural camouflage, and the Squadron pitched its headquarters in a wood next to No 501's Hurricanes, where an army of bugs and mosquitoes made life uncomfortable. After a few local patrols in the morning, eight volunteers were called for to do a bomber escort to the north.

All went well until they reached St Quentin, which was supposed to be well behind the Allied lines, where they were shot at by ack-ack whose accuracy could only be German — as usual, the Hun was not where he was supposed to be. By now their fuel was getting low, so Hilly Brown, who was leading, landed at an airfield to check on the situation. He was told to get airborne immediately, as the Germans were expected to arrive within five minutes. His suspicions were confirmed by a Henschel 126 which appeared over the hill, so he shot it down. He led the Hurricanes to Plivot on weak mixture, where they were refuelled by No 139 Squadron before returning to Anglure at dusk.

Sgt Albonico was missing. He had last been seen diving down to investigate troops on the ground at St Quentin, and was presumed to have been brought down by light flak. He was later reported to be a prisoner of war.

After dark, the pilots crowded into a little cafe in Pleurs which the Bull and Moses had selected as the Officers' Mess. In their not very private surroundings they were told in hushed tones that the Squadron had been tasked for a major escort operation on the following day. Then, after a good meal, they repaired to the grounds of a nearby chateau and pitched their camp under the stars.

19 May 1940

From the very hour of daybreak the pilots were ready, and sat on the grass by their Hurricanes, brooding about the mission to come. These were no longer the laughing, casual boys of Vassincourt — they were tired old men, suddenly and bitterly matured in the ways of battle. But they were still a Squadron, still the finest flying team that the war had yet produced, undeterred and unbroken by the fearful odds and fatigue that they had faced for nine days without rest.

The previous night's briefing had told them that every British bomber still airworthy in France was to be hurled into a last ditch effort to stop the Germans crossing the Aisne — the bridges had not been blown up as planned by the routed French Army. Three fighter squadrons had been detailed to cover the target for 15 minutes each: Nos 73 and 501 were to go in first, and No 1 had been selected for the final period by which time the opposition was expected to have been fully alerted. Such was the price of fame!

At last, at 10.00hrs, they were ordered into the air, and Johnny Walker led the 10 Hurricanes away to the northeast. On the way to the target several unescorted formations of German bombers were sighted, and it broke their hearts to have to ignore them. They reached their patrol line over the Aisne at 20,000ft, but of the German fighters there was no sign — the 15 minutes passed completely without incident. The Hurricanes turned back towards Anglure in a long, gentle dive, overhauling a formation of 25 Blenheims as they did so.

As they came alongside the Blenheims, Walker realised his mistake — they were Heinkels, and completely unescorted. With sudden elation, the Hurricanes attacked individually — Hilly Brown shot one of the Heinkels down in flames and Sgt Soper destroyed two of them, but stopped an incendiary bullet which set his engine on fire. He dived out of the fight so fast that he actually blew the fire out. The cockpit was now completely inundated with glycol, and Soper had to make a wheels-up landing practically blind, knocking himself out in the process.

Paul Richey had an absolute field day. He sent one bomber cartwheeling down on fire and destroyed two more before being badly wounded in the neck by the rear gunner of his last victim. After a terrifying forced landing he was rescued by the French and taken to hospital in Paris. Two further Heinkels were shared by various members of the Squadron, and several more were damaged, before the fuel ran out and No 1 had to return to Anglure.

On landing, the pilots discovered that Halahan had repeated his request to have his experienced men repatriated without delay: they were now so exhausted that their chances of survival in further combat were dwindling fast. In view of the imminent collapse of Allied resistance in France, the CO was quite surprised to find that his wish was granted immediately. On 23 May he handed over No 1 to Sqn Ldr D. A. Pemberton from HQ No 67 Wing and left for England in the company of those who had done so much to write the name of their Squadron across the pages of history — Walker, Hanks, Kilmartin, Stratton, Palmer, Mould, Lewis and Soper.

CHAPTER TEN

On the Run

Sqn Ldr Pemberton now faced the Herculean task of whipping 12 new pilots into shape. All that now remained of the old guard were Hilly Brown — promoted to Flight Lieutenant to take over 'A' Flight — and Sgt Arthur Clowes. Those like Peter Boot, Stavart and Goodman, who had arrived shortly before the invasion of France, now had the combat experience to lead sections on patrol, but the general level of experience had taken a very distinct dive. 'B' Flight was taken over by Flt Lt Fritz Warcup, who arrived on the 20th with two pilot officers from a Spitfire squadron in England, but the rest of the Squadron was fresh from the training schools.

Fritz Warcup was given a real baptism of fire on his first operational trip with the Squadron. Under the supervision of Hilly Brown he led 12 Hurricanes to intercept some Stukas reported to be attacking the BEF near Arras. The deteriorating weather forced the Hurricanes down to 2,000ft over the target area, and although no dive-bombers were seen, they ran headlong into very heavy and accurate flak. Warcup immediately ordered the Squadron to open out and evade. Most of the pilots, under

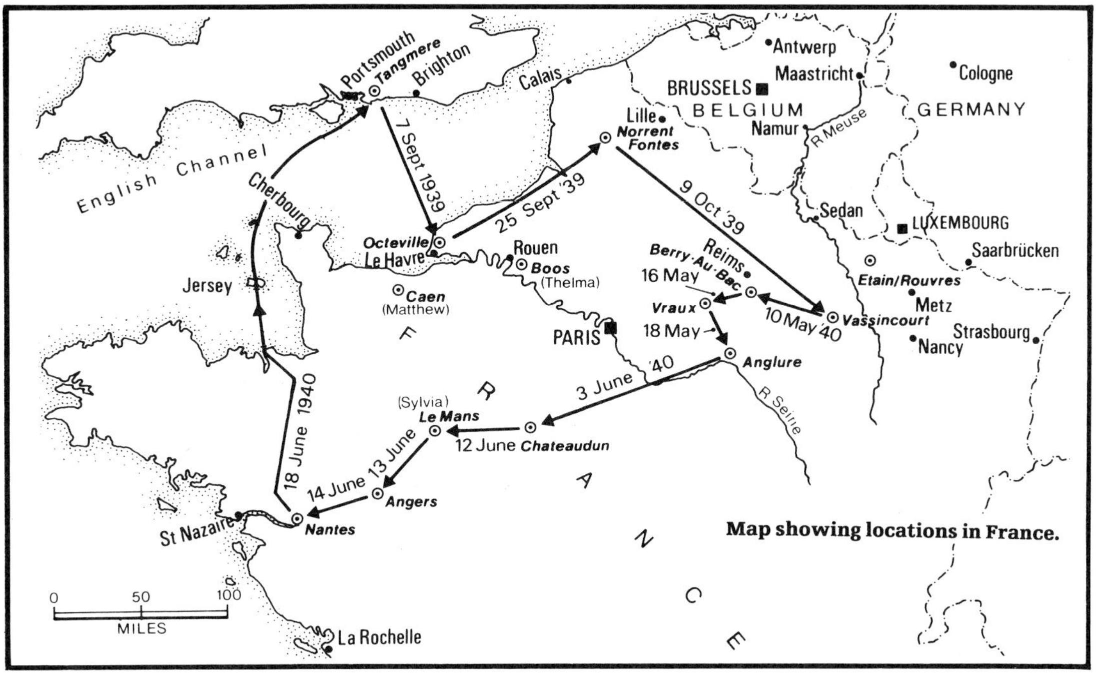

Map showing locations in France.

Above:
Moving back. AC Bill Azzaro (centre) snatches a quick bite during the retreat through France.
W. Azzaro

fire for the first time, wisely sought refuge in the clouds and made their way back to Anglure. The leader, however, under the mistaken impression that he still had 11 Hurricanes somewhere behind him, pressed on with the patrol and got bounced by six Bf109s. They took turns to shoot bits off him until he, too, managed to slip away into the clouds. On the way home he joined up with Sgt Clowes and Fg Off Thom, and shot down a passing Heinkel over Rouen; Plt Off Dibnah destroyed a Dornier 17 at 500ft near Boos airfield at about the same time.

As the German forces were now closing in on the Allied armies at Dunkirk, there was little activity on the part of the Luftwaffe in the Paris area for some days. As a result the Squadron continued to fly patrols further to the north on the Sennesous line and to escort the few remaining British bombers on a variety of missions over enemy-held territory. Fg Off Thom was captured on the 25th. Roland Dibnah recalled: 'Don Thom was the worst navigator ever spawned. His ultimate effort was to put red on blue and fly north from France until he ran out of fuel and was put in the bag by the German army'.

On the following day, Dibnah was wounded in the thigh over Ochamps. He managed to fly 30 miles before force-landing at Nancy, which was still in French hands, and was taken to hospital at Bar-le-Duc where he remembers the agony of having his dressings changed when they were too rushed to soak them off before his hurried evacuation.

By the end of May there was a marked increase in German reconnaissance in the Paris area. Although Anglure's natural camouflage kept its squadrons hidden from prying eyes, Fg Off Salmon was sent off with Moses Demozay to survey possible airfields further south. This proved to be a wise precaution, for on 3 June the Squadron was ordered to move to Chateaudun, leaving behind several brand new Hurricanes that had just been delivered from England. These were collected later by Sqn Ldr Pemberton, together with Plt Off Pat Hancock and two other newly arrived pilots.

At 04.00 on the following morning, Hilly Brown led 11 aircraft to patrol west of Rouen all day. After refuelling at Boos airfield — codenamed 'Thelma' — they sighted a mixed formation of 60 Heinkels and Dorniers at 15,000ft escorted by a similar number of Bf110s. Brown led 'A' Flight in to hold off the escort and ordered Warcup's men to attack the bombers. In the initial attack Plt Off Hillcoat destroyed one of the Heinkels but was forced down with a bullet in the engine. Sgt Arbuthnot had to return to 'Thelma' with his radiator shot away after flying through a formation of 25 enemy aircraft.

'A' Flight meanwhile had tackled a group of 30 Bf110s, shooting down three between them and scattering the rest in disorder before going in to help 'B' Flight. Hilly Brown forced down a Do17 and Peter Matthews destroyed an He111 which spun into the ground from 8,000ft. Peter Boot was also enjoying himself, disabling first a Heinkel and then a Dornier, before running out of ammunition. He then dived aggresively at a straggling Heinkel which, to his great surprise, made an immediate forced landing in a field near Rouen.

For the next five days they patrolled from 'Thelma' and returned each night to Chateaudun, during which time Roland Dibnah was rescued from an ambulance convoy at Le Mans and was flown back in the Magister for a ribald reunion before being evacuated by air to England. It was then learned that the German Army was advancing on Paris from the Somme, and No 1 was ordered to move south once more. On the night of 12 June the groundcrew were ordered to move to 'Sylvia', an airfield near Le Mans. After a hectic night on roads crowded with refugees, they arrived at 'Sylvia' to be greeted with the tale that the Squadron had been diverted to Angers, a further 50 miles to the southwest. Flt Lt Robbie Roberts, the Adjutant, tried to get confirmation of this rumour, but even HQ No 67 Wing, which he discovered in a nearby wood, could throw no light on the subject. He waited all afternoon in vain for the aircraft to arrive, and eventually set off with the convoy for Angers at 18.00hrs. They were passing through Le Chartre an hour later when some Bf110s appeared and started strafing the road. Fortunately Roberts got all the groundcrew into the ditch in time, and no one was hurt. On reporting to Angers aerodrome in the early hours of the 14th, the Adjutant discovered that the Squadron was not there either. He was tempted to return immediately to 'Sylvia', but the troops were completely exhausted and he decided to pitch camp in the grounds of a nearby chateau.

Sqn Ldr Pemberton had meanwhile received orders to move northwest! On the 13th the Hurricanes flew up to Matthew, near Caen, to spend the day patrolling the Seine and covering the evacuation of troops from Le Havre. On the following morning the CO heard that the groundcrew had reached Angers. After further patrolling of the Channel ports the Squadron was ordered to join up at the final evacuation centre at Nantes.

Darkie Clowes led the last section out of Matthew, shooting down a Bf109 en route to Angers, where he informed Roberts of the latest moves before flying down the Loire to join the CO at Nantes.

No 1 arrived at Nantes to find the airfield looking like several Empire Air Days all at once, covered from end to end with Hurricanes, Battles, Blenheims and Magisters in varying states of repair. No 67 Wing had arrived at the same time and dispersed with the Squadron into a nearby wood alongside the Hurricanes of Nos 73 and 242 Squadrons. There were a number of brand-new Hurris on the field and some of these were issued to No 1 to replace aircraft damaged on recent patrols. Every available anti-aircraft gun had been collected and these were placed around the dispersals in event of attack by the Luftwaffe. Hilly Brown and Fritz Warcup had gone off on patrol together soon after lunch and were now well overdue; by dusk, the CO was getting very worried by the absence of both his Flight Commanders, especially as all units were standing by to move on — the chances that they had diverted were waning as time passed. Then came the news that Paris had fallen, which struck a severe blow to the morale of the French troops on the airfield, including Moses Demozay.

By the morning of the 16th it was obvious that everyone would be leaving for England very soon. Flt Lt Barber, the Squadron's new Intelligence Officer, and 'Knackers' Henniker, the tireless Rolls-Royce rep, were detailed to lead the main party away at dawn to the docks at nearby St Nazaire. Taking only what kit they

Groundcrew and pilot attempt to start a reluctant Hurricane. *Squadron Archives*

could carry, they were evacuated at 11.30 via Brest in a tramp steamer bound for Plymouth.

No 1 Squadron flew its last operational patrol in France on the 17th. Sgt Bill Berry was leading a section of three aircraft over the St Nazaire docks when a formation of Heinkels swept in at low level to attack the embarking troopships. Berry frantically led his section into the attack, but was too late to prevent the leading enemy from releasing its bombs on the *Lancastrian*, a converted liner with over 4,000 troops already aboard. The pilots were horrified to see the ship erupt in flames: it later sank with very heavy casualties. It was small consolation that the bomber that struck the ship was sent blazing into the river by Sgt Berry.

Activities back at Nantes were hotting up all day. Arrangements were being made for the remainder of the Squadron to move down to La Rochelle, where the rest of the groundcrew were supposed to embark but from where the Hurricanes could not reach England — it was rumoured that they were to fly home via the South of France and North Africa. Fortunately, this order had been rescinded by the following day, although all but 16 of the men were sent to La Rochelle for final embarkation. At midnight of the 17th, two officers and 42 men were bundled into two diminutive colliers, and said goodbye to France after a stay of nine long months.

All that now remained of No 1 Squadron in France was 12 Hurricanes, their pilots, 16 groundcrew to start them up, and one faithful interpreter to wave them farewell; Moses had been ordered to report to the crumbling French HQ in Nantes, where he faced a highly unpromising future. After two early morning patrols over Nantes and St Nazaire, the pilots were ordered by Sqn Ldr Pemberton to start up at 11.45. After take-off they leaned out their mixtures and settled down for the long flight back to England, landing to refuel at Jersey and Weymouth before reaching Tangmere for a late lunch.

Moses Demozay sat deep in thought as he watched them fly away. The 16 groundcrew faced a difficult trip to La Rochelle in the last of the Squadron's weary trucks, and the situation on the roads was getting worse by the hour. Yet there on the airfield stood a Bristol Bombay, abandoned with a broken tail-wheel but brim-full with fuel. Would they, he asked, risk it with him?

Within minutes a couple of riggers had made a makeshift repair to the tailwheel and Moses was sitting in the cockpit deciphering the unfamiliar controls. A sergeant operated the throttles while the Frenchman coaxed the strange machine into the air. They crossed the Channel without incident, and Moses started thinking about somewhere to land. He chose Tangmere initially, but the NCO told him that most of the boys on board lived in East Anglia, and would he like to put down at Sutton Bridge? Moses, always the gentleman, immediately agreed, and in due course set them down not too far from home.

The first person he met at Sutton Bridge was Billy Drake, who had just arrived at the station from hospital. Billy told Moses that Bull Halahan and the others had all been posted to instruct at a Hurricane OTU at Aston Down in the Cotswolds. Moses thought that this would be just the place for him to get back to full time flying. Billy lent him some money, and he set off without delay to rejoin his old friends — No 1 Squadron had not heard the last of Moses Demozay.

Hilly Brown was waiting for them at Tangmere. He had baled out over France and had managed to get a lift to Brest where he caught a boat to Southampton. He had no idea of what had happened to Fritz Warcup, who had now, like Brown, been posted missing: Warcup was later reported to be a prisoner of war.

So ended the Battle of France. No 1 Squadron, which had been the first British unit to cross the Channel to war, was one of the very last to leave. Its endeavours and success did not go unrecognised. As the summer of 1940 built up to its fateful climax over the fields of southern England, King George announced a unique block award of decorations for gallantry in France. This read as follows:

The Distinguished Flying Cross
Sqn Ldr Patrick John Handy Halahan
Flt Lt Peter Russell Walker
Flt Lt Peter Prosser Hanks
Flt Lt Mark Henry Brown
Fg Off John Ignatius Kilmartin
Fg Off Leslie Redford Clisby
Fg Off Paul Henry Mills Richey
Fg Off Peter Mould
Fg Off Cyril Dampier Palmer
Plt Off William Hector Stratton

The Distinguished Flying Medal
FS Francis Joseph Soper
Sgt Arthur Victor Clowes
Sgt George Frederick Berry

CHAPTER ELEVEN

The Battle of Britain

After refuelling at Tangmere, the pilots of No 1 Squadron were ordered on to Northolt, on the western fringe of London. Plt Off Pat Hancock remembered that they received a poor welcome at the Squadron's old home: 'We did not have any rest at Tangmere, only lunch, and we were regarded with some disfavour as we ponged to high heaven. Washing had been somewhat skimpy in the last days in France'.

At Northolt they were reunited with their vagrant groundcrew and started to bring the Squadron back to full fighting trim. The establishment was cut down from 300 to 225 men and the ground equipment lost in France was replaced. The loss of most of the Squadron's documents was particularly troublesome, the Orderly Room staff spending many long hours filling in the gaps.

Fg Off Brian Hillcoat was promoted to take command of 'B' Flight and Plt Offs Davey and Elkington were posted in from an Operational Training Unit. Flt Lt Roberts, who had worked so hard as Adjutant in France, was posted out to train as a Technical Officer. He was replaced by Plt Off L. W. Grumbley, who had retired as a Major from the Indian Army some years before the war after 40 years of service.

Most of the tired old Hurricanes from France were replaced with new machines, and under the guidance of the CO and Hilly Brown, the Squadron embarked on a period of intensive training by day and night which lasted until the beginning of August. No 1 was declared operational once more on 3 July, and flew a number of uneventful patrols until 19 July, when three aircraft led by Sqn Ldr Pemberton intercepted a Heinkel over the South Downs. They exhausted their ammunition in a series of unsuccessful attacks, and Plt Off Browne ('Browne with an E' — to distinguish him from Hilly) was shot down while attempting to trail the raider and advise other units of its position. He crash landed near Brighton and managed to crawl clear of his Hurricane before the fuel tanks blew up. To add insult to injury, the Heinkel was later brought down by No 43 Squadron. The Squadron was cheered up, however, by the return of Roland Dibnah from hospital.

No 1 was able to get its own back on the 'Fighting Cocks' when the Squadron was detached to Tangmere from the 23rd. After two days at its old peacetime station the Squadron recorded its first successful action in English skies. Portland had been bombed on the morning of the 25th, and several patrols were flown in the afternoon to cover shipping outside the harbour. Hillcoat was on patrol with Sammy Salmon and Randy Goodman at 15.20hrs when they were attacked by six Bf109s. In the ensuing dogfight Salmon sent one of the enemy down in a spin from which it could not have recovered, although it was not seen to crash. Goodman had a lucky escape when another of the Bf109s came in on a beam attack, pulled its turn too tightly, flicked, and spun into the sea.

While patrolling the Isle of Wight on 31 July, a section led by Peter Matthews damaged a Dornier 17 after expending the prodigious total of 5,000 rounds of ammunition. As his Squadron's marksmanship had obviously deteriorated since the Battle of France, Pemberton was quick to arrange a bomber affiliation exercise when No 1 returned to Northolt on 1 August. Six Blenheims turned up a few days later to act as targets for their simulated attacks, and a valuable day's

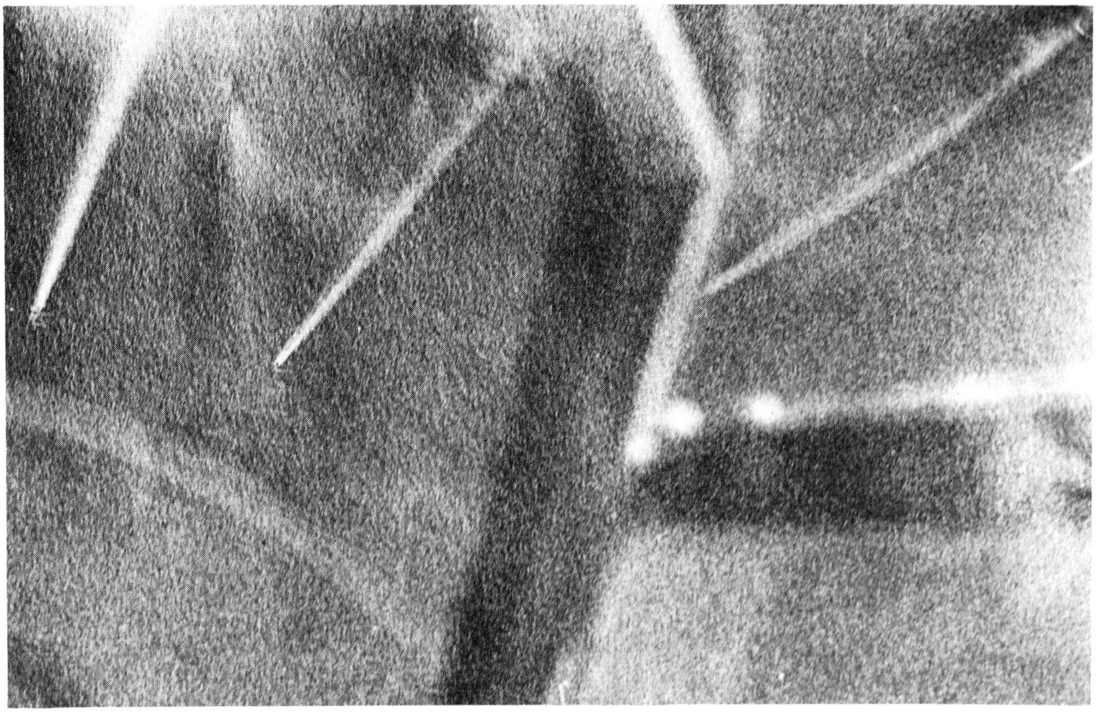

Top:
Three of No 1's aircraft on patrol during the Battle of Britain. This shot was taken through the windscreen of Plt Off J. F. D. Elkington's Hurricane.
Wg Cdr J. F. D. Elkington

Above:
Plt Off Elkington's Hurricane closing on three vapour trails during the Battle of Britain.
Wg Cdr J. F. D. Elkington

training resulted. Unfortunately one of the bombers crashed with the loss of its crew during the exercise while taking evasive action at low level.

So far the attacks on the Channel convoys, ports and coastal airfields had merely been probing manoeuvres designed to draw up the RAF's fighters, and thin them down before the final blow fell on the main sector airfields. But by 11 August Goering's forces were ready, and a fleet of several thousand German combat aircraft stood poised to unleash the Blitzkrieg on England.

On the morning of the 11th, three sections of No 1 Squadron were detached to Tangmere to spend the day patrolling the south coast. They were sent out over the Channel at 16.00, and when well out to sea were attacked by a formation of Bf110s. After a long struggle the Hurricanes managed to get the upper hand, and one of the enemy aircraft was destroyed in flames by Hilly Brown. Plt Offs Chetham and Goodman attacked and damaged another of the Messerschmitts, which then let out a drogue on several hundred feet of wire. This did not inconvenience either pilot, and Chetham managed to silence the rear gunner and set the starboard engine on fire. Goodman then closed to within 50yd of its port wing and blew the Messerschmitt apart with a five-second burst. Plt Off Davey's aircraft was damaged in the fight, and he was killed while attempting a forced landing on the Isle of Wight. He had only recently joined the Squadron.

During the next three days, sections of the Squadron were detached to Hawkinge, which was bombed while they were there, and to North Weald and Manston, but no successful interceptions were made.

Goering had been forced by cloudy weather to postpone his major blow on the sector airfields until 15 August, when he sent out 1,790 sorties to pound Fighter Command. No 1 Squadron was not involved in the battles over Kent during the morning, but was sent to North Weald at midday to patrol the Essex coast.

At 14.30hrs, nine Hurricanes were scrambled from North Weald to intercept a raid that was reported to be approaching Harwich at 1,500ft. Hilly Brown sighted the raiders 10 miles out to sea from Harwich harbour, the enemy force consisting of 20 Bf110s armed with 250kg bombs, escorted by a *Staffel* of Bf109s. Brown did not know it at the time, but he had encountered *Erprobungsgruppe 210*, a crack trials unit of the Luftwaffe which had been hand-picked to pioneer fighter-bomber techniques. Before long the Hurricanes were in trouble. Hilly Brown was attacked by a section of Bf109s and was wounded in the face and hands; he baled out, and was rescued by a passing trawler which landed him at Harwich. In a running fight all the way to Martlesham Heath, which the Germans had been sent to attack, Plt Off Browne and Sgt Shanahan were shot down and killed. Over the target the Bf110s were engaged by a section led by Fg Off Peter Matthews, who attacked three of the enemy, and one of them was sent down with an engine on fire. Plt Off H. T. Mann was attacked by one of the Bf110s which had released its bombs, but he managed to evade it; he was then engaged by a Bf109, but manoeuvred on to its tail and blew the top off its cockpit. By this time he was very short of fuel, and had to land at Martlesham airfield, which had been heavily bombed during the raid. Plt Off Elkington still had some petrol left and pursued the Germans out to sea. He was attacked by a Bf109 but managed to get on to its tail. After three short bursts the Bf109 turned slowly on to its back and dropped like a plummet into the North Sea.

Sqn Ldr Hilly Brown joined the Squadron from Canada in 1937. After the Battles of France and Britain, he was promoted to command No 1 when Sqn Ldr Pemberton was killed at Wittering in October 1940. He led the unit on its early sweeps back into France, but was later shot down when leading a Spitfire wing over Sicily, with 17 kills to his credit. The Luftwaffe dropped a wreath over Malta to say that he had been buried with full honours. *IWM*

The atmosphere at Northolt that night was not a happy one. None of the pilots felt like going up to London after duty, for the loss of three of their number in five days seemed to tell them that the Squadron's good luck was beginning to run out. When morale started to flag there was plenty to grouse about. Although most of the boys had lost all their personal kit in France, the Air Ministry was still arguing over the £60 replacement allowance due to each man. It had even been wryly suggested that the authorities were trying to save money by waiting until they had all been killed before authorising the payment.

At midday on 16 August No 1 was ordered to patrol Tangmere at low level. It fought a minor skirmish with some Bf109s, but nothing was claimed. Plt Off Elkington was forced to bale out when his Hurricane was hit over Thorney Island.

At 16.30hrs, Sqn Ldr Pemberton led the six Hurricanes of 'A' Flight from Northolt to patrol Tangmere again, followed by Hillcoat's six aircraft which took off five minutes later. The pilots arrived over Tangmere to discover that their ancestral home had been badly bombed while they had been refuelling and resting, and then they were diverted back towards London to intercept another big raid. Near Guildford they were alerted by a call of 'Tally-Ho!' from their Squadron Commander: approaching them were 120 Heinkels in three huge stepped-up waves, closely escorted by Bf110s and followed by a further formation of Junkers 88s.

Pemberton gathered his forces, climbed for height and attacked the first wave of bombers from the beam. His first burst was delivered with deadly accuracy and sent a Heinkel down on fire. Recovering from his attack, he was hit in the engine and had to leave the fight. Peter Matthews took over the leading section and attacked a formation of the escort fighters. He managed to shoot one down, but Peter Boot was hit in the glycol tank and limped away to make a forced landing on the Hogsback. His Hurricane was totally destroyed when it ploughed through a forest of sharpened anti-invasion stakes, but Boot escaped unhurt, shaken and seething with unreasoned anger at the local defence expert who had planted them.

Darkie Clowes attacked a Heinkel which he saw go down in a vertical dive, shot at another Heinkel without visible result, and then dropped below cloud on the tail of a Junkers 88, which he damaged with the last of his ammunition. Four Spitfires then appeared on the scene, and the German aircraft dived down and crash-landed without waiting for their fire. Meanwhile, Randy Goodman picked out a Heinkel after the initial Squadron attack, and sent it flaming to the ground near Petworth. While he was circling the crash at 500ft, the Heinkel's fuel tanks blew up and damaged his Hurricane's engine. He beat a hasty retreat to Northolt.

Hillcoat's 'B' Flight was also in the thick of the battle. Bill Berry saw two of his targets break away from the formation, one of them diving vertically down, obviously out of control. Sammy Salmon had been left behind during the initial attack, and climbed to a great height to dive on the bombers from behind: as he did so, a Bf110 flew into his view, so he shot its port engine to pieces and sent it down in flames.

Plt Off Red Stavart had attacked the bombers head on with a section of 'B' Flight. He made a frontal attack on a Heinkel which promptly dived into the clouds, so he rolled on to his back and pulled the Hurricane through the second half of a loop. He broke cloud in time to see his victim explode on the Downs beneath him.

With the depression of the previous evening washed completely from their minds, the pilots of No 1 Squadron made their individual and victorious ways back to Northolt. In the space of one hour they had completely disrupted a major raid and destroyed four He111s, one Ju88 and one Bf110, plus two Heinkels and one Messerschmitt probably destroyed. Their only regret was that they had missed out on the earlier raid on Tangmere, in which they were delighted to hear that No 43 had claimed 17 Stukas destroyed.

The following day was fairly quiet, but both Flights were deployed to North Weald on the 18th. At 13.15hrs a raid was reported to be approaching Rochford, and the Squadron was scrambled to intercept over Southend. At 2,000ft it encountered 12 Bf109s coming out of cloud, but the enemy attacked without success, for Pemberton split the Squadron to bring Hillcoat's Hurricanes round behind the Messerschmitts. On seeing this tactic the enemy scattered, and the CO closed in behind one of the stragglers. He fired two bursts from about 250yd but the German continued to dive across the Thames Estuary and hedge-hopped away across Kent. He gave it one last burst as it drew out of range and was surprised to see it blow up and crash near Tenterden. There were large numbers of enemy aircraft over the Estuary at the time, but heavy cloud prevented any more successful engagements.

Having returned to Northolt, No 1 was scrambled again to patrol overhead at 10,000ft. After 30min nothing had turned up and the Squadron was ordered to fly south and patrol over Shoreham. On the way it ran into 60 unescorted Dorniers over the South Downs.

Fuel was getting very low by this time, but Hillcoat, Stavart and Goodman made a beam attack which shot down one of the bombers, the crew baling out over the coast. Goodman was examining one of the parachutes when he saw a Bf110 heading for home at very low altitude with one engine smoking. After a careful look at his petrol gauges, Goodman dived on the Messerschmitt and took a perverse delight in putting its remaining engine out of action. He just had time to watch it ditch before returning to base with dry tanks.

Plt Off Freddie Birch was detailed for a solo patrol on the following night. Having thoroughly lost himself, he had the misfortune to stray into the London balloon barrage, and baled out when his aircraft struck a cable. He landed with very minor injuries on a rooftop in Finchley and then spent some time convincing the locals that he was British. He was eventually persuaded to make another terrifying jump from the eaves into the arms of a burly fireman standing on top of a turntable ladder some 5ft away.

Throughout the following week the Squadron patrolled without success. The weather was unsuitable for the Luftwaffe to come over in full strength and No 1 never seemed to be in the right place at the right time. However, the comparative lull enabled the CO to organise some very necessary flying training for his newer pilots, and it gave the Squadron some badly needed rest.

The weather improved again on the 28th, and 11 Hurricanes left Northolt to patrol the Hornchurch sector at 12.30hrs. When over Rochford at 15,000ft they met 30 Dorniers closely escorted by a dozen Bf109s. In the attack that followed one of the Dorniers crashed on Rochford aerodrome and another was seen to plunge into the sea near the Tongue lightship. A third was badly damaged and was last seen heading out to sea at low level, closely protected by its fellows. The escort fighters had taken little part in the

No 1 Squadron was moved north to rest on 7 September 1940, having been continuously in action since the invasion of France. Here Sgt Darkie Clowes climbs into the aircraft he flew throughout the Battle of Britain before a quieter patrol from Collyweston, a satellite airfield of the Squadron's present base at Wittering. *Squadron Archives*

battle. It turned out that the heavy losses of the past 10 days had caused Goering to order his fighters to stay close to their charges, from which position they were unable to do their job properly.

Two days later the Luftwaffe launched 1,500 sorties over Southern England. The losses of the previous fortnight meant that only 500 of these were bombers, covered by every serviceable fighter that the Germans could scratch together.

At 16.30 No 1 was scrambled to intercept a raid over Essex. While passing north of London the pilots saw six bombers being attacked by another Hurricane unit, and joined in from behind with cries of glee. As Sqn Ldr Pemberton was about to open fire he realised with horror that his target was a Blenheim — the other Hurricane squadron was practising bomber affiliation in the battle area! At the last moment he managed to call his pilots off and tragedy was narrowly averted.

As they formed up again to head eastward, an untidy formation of about 30 Heinkels appeared. They were escorted by an even scrappier group of Bf110s, and it appeared that they had already been in a fight. The enemy aircraft were so widely spread out that a Squadron attack was impractical, so the 11 Hurricanes went in individually.

Plt Off Birch damaged a Heinkel with his first burst and Plt Off Hancock fired at another which he set on fire. Sgt Clowes damaged a Heinkel and then a Bf110. Peter Matthews picked a straggling bomber, but ran out of ammunition before he could finish it off; he saw a Hurricane of No 56 Squadron come in and send it down.

By now the fight was approaching the East Coast. Plt Off Mann managed to turn the tables on a couple of attacking Bf110s, outmanoeuvring them and scoring hits on both, and just before the Hurricanes had to turn for home, Sgt H. J. Merchant got on the tail of a Heinkel and sent it crashing into a cemetery in Shoeburyness. Every aircraft returned to Northolt without a scratch.

On the following morning the Squadron was at readiness at dawn. So far Northolt had escaped without a serious raid, but Debden, Hornchurch, and Biggin Hill particularly, had taken a fearful pounding on the previous day.

They were scrambled at 07.40 and ordered to patrol Chelmsford. On arriving over the town they met a mixed formation of over 100 Dornier 17zs and Junkers 88s, escorted by a flock of Bf110s so large that they did not bother to try to count them. With nothing but grim determination, the seven Hurricane pilots pulled their boost plugs, climbed into the protection of the morning sun and swung back into the heart of the German raid. One group of Bf110s had seen them coming and swept down to try to cut them off.

Sgt Merchant, flying No 2 to the CO, was diving on the bombers with his leader when the Messerschmitts attacked. One of them, trying to fasten on to Pemberton's tail, flew right across Merchant's nose. He fired instinctively as it passed, and it blew up right in front of him. A moment later he fired at another Bf110 and damaged it. Just then he felt cannon shells thudding into his rear fuselage and the Hurricane burst into flames. As the heat seared his face and hands, Merchant tugged frantically at his hood release. It gave way and he dived clear of the blazing aircraft. After parachuting to safety, he was taken to Halstead Cottage hospital suffering from burns.

Sqn Ldr Pemberton had meanwhile escaped the attentions of the Messerschmitts and severely damaged a Ju88. Darky Clowes made a frontal attack on a formation of 15 Dorniers, spraying them liberally with his eight Brownings. He saw two of them break away, and one of these spun down enveloped in smoke. He was then jumped by a bunch of Bf110s and looped tightly into their attack. As they met head-on he fired a short burst, which smashed into the starboard engine of one of his assailants. Peter Boot and Roland Dibnah also damaged a Bf110 apiece then had to divert to Hornchurch, short of fuel. 'Dib' failed to find the aerodrome in the morning mist and landed in a field to ask the way. 'Over there', pointed a farmer. Dibnah didn't have time to raise his wheels before crossing the boundary of Hornchurch.

At 10.45hrs on 1 September, Flt Lt Hillcoat led the nine serviceable Hurricanes to patrol Biggin Hill, which had been badly hit again on the previous day. Before long they were vectored on to a raid approaching from the southeast, which they intercepted over Tonbridge. The bombers were already being attacked by Spitfires, and Hillcoat was able to surprise an escorting formation of Bf109s. With the advantage of height and the sun, Hillcoat, Boot, Birch and Chetham each claimed one of the German fighters, and the Hurricanes flew home without a scratch.

On the afternoon of the following day, Hillcoat was leading the Squadron again. This time the Hurricanes met a formation of Bf109s flying alone over Surrey and once more they

Comrades in arms at Wittering in November 1940. Top: Darkie Clowes, Sgts Zavoral and Prihoda (Czech), Plt Off Elkington. Middle: Moses Demozay, Sgts Kuttelwascher, Novak and Stefan (Czech). Bottom: Plt Off Hancock, Hilly Brown, Plt Off Chetham, Sgt Plasil (Czech). *Squadron Archives*

achieved complete surprise, Hillcoat claiming their leader as probably destroyed. Chetham damaged another, but Flt Sgt Bill Berry had to bale out, and was killed when his parachute caught fire. Roland Dibnah, for one, remembered feeling an enormous wrench at the loss of this experienced and immensely popular pilot.

Despite the loss of Berry, No 1 was fighting on with most of its original pilots. Once again the traditions of close teamwork, constant training and outstanding leadership were pulling the Squadron through. By this stage in the battle No 1 Squadron was the most intact unit in No 11 Group. Each pilot, regardless of his fatigue, had flown an average of two operational sorties per day over the past few weeks; the CO, in particular, had not spared himself. As well as the routine administrative work he had to do every day, Sqn Ldr Pemberton flew 66 sorties during August, and was, like the rest of his Squadron, a very tired man.

A few nights previously a series of minor attacks on Berlin by British bombers had caused Hitler to seek revenge on London. By 6 September the attacks on the sector airfields had virtually ceased and the full weight of the Luftwaffe was turned against the capital. Now Fighter Command, its resources stretched to the limit, was able to regroup and fight back with fresh squadrons rotated from the North to replace those hardest hit. The shield, though dented, was far from being broken.

On 6 September Hilly Brown led nine Hurricanes to patrol Kenley at 20,000ft. When just over the patrol line they engaged a fast-moving formation of Ju88s with a heavy escort of Bf110s. A series of individual combats took place, in which Brown and Stavart each destroyed a Ju88. Dibnah and Goodman took on the escort and both claimed Messerschmitts, but Goodman was in turn attacked and had to bale out without injury. Dibnah recalled:

'It was a horrifying situation of being trapped into a head-on attack with the 110, and killing him before he killed me. Then I remember a lone 109, pitty-patting back to France, hoping no one would see him; the careful stalk in his blind area, head turning like an owl, not believing it can be this easy, pulling up at less than 100 feet and firing; instant reaction, hood flying off, bailing out through my bullet stream, falling 17,000 feet into the channel, chute unopened'.

At 13.00hrs the Squadron was scrambled to intercept a heavy raid approaching London from Dungeness. Hilly Brown had difficulty in getting his engine started and so Sgt Clowes led the remaining aircraft to the scene. While climbing to intercept the bombers the Squadron was attacked by a large formation of Bf109s, but Clowes fought a brilliant defensive action and the Hurricanes escaped unscathed. During the dogfight Stavart fired at a Bf109 which was last seen diving for the coast with black smoke pouring out of it.

Two new Hurricanes were received from the factory during the afternoon. The groundcrew, working far into the night as usual, produced 11 serviceable aircraft on the morning of the 7th. London's blazing skyline had suddenly given them even more to work for.

After two uneventful patrols in the morning, the Squadron spent the afternoon of the 7th at cockpit readiness. By 17.00hrs 300 German bombers escorted by 600 fighters had penetrated the Thames Estuary, and headed for London in a series of massive waves. All the way from North Foreland they were attacked by squadron after squadron of Spitfires and Hurricanes, but still they came on.

At 16.25 the 11 Hurricanes had scrambled to patrol northeast London, led by Hilly Brown, and on reaching the area they saw a large formation of bombers escorted by about 100 Bf110s. Sgt Clowes destroyed one of the escorts in the fight that followed, while Flt Lt Holderness, who had just joined the Squadron, shot down a Dornier 17z.

The aircraft returned to Northolt without loss by 17.30hrs, and were immediately refuelled and rearmed. At 17.50 they were scrambled again, this time under the command of Sqn Ldr Pemberton. A further large wave of Dorniers was encountered and two of them were badly damaged by other new arrivals, Flt Lt Finnis and Plt Off Homer.

By now pilots and groundcrew were suffering very badly from fatigue, but the Squadron was still flying as a team and had suffered surprisingly few casualties. Fortunately it was now No 1's turn to be pulled out of the front line and transferred to a quieter section in No 12 Group. The move meant that the pilots missed the action over London when the Battle of Britain reached its dramatic climax on 15 September, but in their state of exhaustion it was the wisest decision.

No 1 Squadron was posted to Wittering on 9 September and was dispersed to the satellite airfield at Collyweston on arrival. The CO was now able to send some of his men on leave, and the Hurricanes were grounded one by one for repair, overhaul and the fitting of VHF radio sets.

From 13 September single aircraft of the Squadron began to patrol the Wittering sector at night, but, without the help of local radar stations' equipment, no success was achieved for some time. Day patrols were also carried out for the rest of the month, but the Luftwaffe was fully committed further south and nothing was seen.

On the 13th, Sqn Ldr Pemberton and Plt Off Boot were awarded the DFC for gallantry displayed in operations in France and from Northolt, while Sgt Arthur Clowes was granted a well-deserved commission as a result of the leadership he had shown during the previous year.

Many of the pilots who had flown with No 1 Squadron throughout the Battle of Britain were posted out during October, most of them to become instructors at the fighter schools. Their places were taken by a large influx of Czechoslovakian pilots from No 6 OTU. These exiles from an occupied country were filled with a remorseless revenge for the Germans. Names like Velebnovsky, Kuttelwascher and Dygryn began to appear on the posting notices, and before long the nominal roll read like the bottom line of an oculist's test chart.

Not all the new arrivals were Czechs. On 16 October, Lt Moses Demozay returned to No 1 Squadron. The former interpreter had completed his conversion to Hurricanes and had pulled every string he knew to return to his old unit as a pilot. Because he still had a family in France, he assumed the nom-de-guerre of Morlaix.

October was a month of very intensive training. Under the command of Plt Off A. H. Pooley, the groundcrew produced a remarkable level of serviceability, and as the weather was also good throughout the month the squadron achieved 1,003 flying hours. Victories against lone Dorniers were scored on the 24th, 27th and 29th by Hilly Brown, Randy Goodman and a section of 'B' Flight led by Sgt Page.

At the end of the month Sqn Ldr Pemberton was congratulated by the AOC No 12 Group for producing such a successful month's flying. Tragically, he did not live long enough to celebrate his Squadron's success. At 07.30hrs on 3 November, the man who had led No 1 through the Battle of Britain, crashed soon after taking off from Collyweston and was killed. Acting command of the Squadron was given to Hilly Brown. A week later the tough little Canadian was promoted to Squadron Leader and his position as CO was confirmed by the AOC. Brown had joined No 1 in 1937 as a Pilot Officer and had been with the unit ever since — a chain of promotion probably unique in the history of the RAF.

The two other remaining pilots who had been with the Squadron in France, Darkie Clowes and Peter Matthews, were promoted to Flight Lieutenant to take over 'A' and 'B' Flights respectively. The promotion of Clowes was in fact so rapid that he never actually held the rank of Flying Officer — it took him just six weeks to progress from Sergeant to Flight Lieutenant.

The training of new pilots continued at Wittering for a further six weeks. On 15 December the Squadron was moved back to Northolt to resume its place in the front line, and for the remaining two weeks of the year No 1 Squadron carried out patrols over Kent and London by day and night without incident. No enemy aircraft were sighted, although Sqn Ldr Brown reported seeing an oil tanker blow up in the Thames estuary on the 19th. The first indication of this was that the clouds boiled up and parted, followed by a terrific shock-wave which rocked the CO's Hurricane.

The New Year Honours List was announced on the 31st. Fg Off Donald Hills was awarded the MBE for his work as the Squadron's Equipment Officer in France, while Brian Hillcoat and Fritz Warcup were mentioned in dispatches. The groundcrew were not forgotten. Flt Sgt Fuller, Sgt Rognard and Cpls Dimmer, Maloney and Trim were all commended for their meritorious service.

On New Year's Eve, Flt Lt Clowes was briefed for a special operation for the following day. He was ordered to take three pilots on a 'Mosquito Raid' to attack coastal defences between Calais and Dunkirk. Six months after they had been driven out of France, pilots of No 1 Squadron would begin to hit back, on the first day of 1941.

No 1 on patrol from Wittering with the Spitfires of No 266 Squadron. *Squadron Archives*

CHAPTER TWELVE

Night Harvest

At 11.00hrs on 1 January 1941, Flt Lt Clowes crossed the English coast with Plt Offs Lewis and Kershaw to perform No 1 Squadron's first offensive mission of the war. Hampered by heavy cloud at 1,000ft, Clowes split up the formation. For the next 20min the Hurricanes strafed German installations between Calais and Boulogne, unopposed by enemy fighters.

During the following months the Squadron was tasked for many such hit-and-run operations. When performed in Flight or Squadron strength, these sorties were known as 'Rhubarbs'. On other occasions a small force of light bombers would cross the Channel with an escort of several squadrons of British fighters to try to draw up the Bf109s. These 'Circus' operations frequently achieved their aim, but often resulted in heavy losses for the RAF.

The newly-promoted Flt Lt 'Darkie' Clowes took over 'A' Flight in November 1940, and never actually held the rank of Flying Officer! Each stripe on the tail of his wasp represents an enemy aircraft destroyed. *Wg Cdr J. F. D. Elkington*

On 4 January No 1 Squadron moved to Kenley where it joined No 615 in forming a Hurricane Wing led by Wg Cdr Johnnie Peel, who had commanded No 145 Squadron with considerable distinction during the Battle of Britain.

The Squadron took part in several 'Circus' sweeps during early February. Plt Off Lewis was lost on 5 February, after he baled out into the Channel, but the score was evened on the 8th when the CO shot down a Bf109 over Quercamps airfield. A number of convoy escort patrols too were flown during February and March. These were sometimes heavily opposed by German fighters, and Plt Off Tony Kershaw was shot down and killed by a Bf109 on 19 March. On the 24th 2nd Lt Moses Demozay led three Hurricanes on a convoy escort off Hastings, just beneath a thin cloud layer at 6,000ft. During the patrol a number of Bf109s kept dropping down singly out of the clouds for a surprise attack, but were chased off by the Hurricanes each time. Moses eventually tired of this and snapped up through the cloud layer to see what was going on. He broke cloud just behind a close formation of the enemy and made a climbing attack on the rearmost Messerschmitt. After a 12-second burst of fire the German aircraft spun into cloud and crashed into the sea.

During March the Squadron was re-equipped with the Hurricane Mk II. This new version was powered by the Merlin XX engine, which gave it an impressive gain in top speed and rate of climb. Additionally, four of the new aircraft were fitted experimentally with 12 Browning machine-guns.

Over the winter months there had been a lot of discussion and rethinking of tactical formation flying. The prewar close attack formations which had been discarded by No 1 Squadron in France were abandoned by most other squadrons during the Battle of Britain, and new tactics were accordingly developed during the quieter days that followed. The main change was the adoption of a pair of aircraft as the basic fighting unit, with the leader of the pair responsible for its positioning within the squadron formation, while his tail was covered by his wingman, or No 2. Sqn Ldr Brown had retrained No 1 Squadron along these lines during the winter, and the new tactics proved particularly suitable for the low level offensive sweeps over France.

Wg Cdr Peel led five of the Squadron's Hurricanes to attack Berck airfield, near Le Touquet, on 7 April. He strafed the field from 200ft in line abreast with two of the Czech pilots, attacking three Bf109s which were preparing to scramble. Peel set two of them on fire, Sgt Stefan destroyed the third and Sgt Goy mowed down a group of a dozen men running towards the machines. Further up the coast, they attacked and damaged several vehicles in a large motor convoy.

On the following morning Fg Off Robinson led three of the Czechs to patrol Dungeness at 28,000ft, and when three black Bf109s were sighted in mid-Channel Robinson was unable to restrain his men. The Hurricanes gave chase to the French coast and a dogfight ensued. Sgt Kuttelwascher stalked a Bf109 which had attacked Sgt Plasil unsuccessfully, Kuttelwascher closing to 30yd before opening fire. The Bf109 literally fell apart and the pieces scattered over a wood near Cap Gris Nez.

A patrol of six aircraft intercepted four high-flying Bf109s over Maidstone on 21 April, the additional power of the new Merlins enabling Fg Off Robinson to climb to 30,000ft and reach the Messerschmitts. The leader damaged one of the enemy who got away by diving fast into the evening sun. Sgt Prihoda was attacked head-on by another Bf109, so he rolled under the German's belly and turned steeply to port, in time to see his assailant diving away towards the coast. He then gave chase with typical Czech tenacity, and caught up with the Messerschmitt over the Channel. After a short burst from the 12 Brownings, the enemy aircraft started pouring out black and white smoke — it was last seen spinning into cloud at about 7,000ft and was claimed as a probable kill.

A week later Prihoda was bounced by another Bf109 near Dungeness. Although it shot him down he baled out and landed safely in a marsh at New Romney.

On 29 April both Flight Commanders of No 1 Squadron, Peter Matthews and Darky Clowes, were awarded the DFC for their consistently fine performances throughout the preceding year. It was unfortunate for their unit that their posting notices arrived on the same day, being both sent to Operational Training Units to pass on their skills to the new generation. They were replaced by Flt Lts K. C. Jackman and W. H. Sizer.

During the following week the turnover in command became complete when Hilly Brown handed over No 1 Squadron to Sqn Ldr R. E. P. Brooker. Richard Brooker had flown with No 56 Squadron during the Battle of Britain and had just completed his rest tour as a gunnery instructor at Sutton Bridge.

Sqn Ldr Hilly Brown left No 1 to command an OTU at Grangemouth, was promoted to Wing Commander and was awarded the Bar to his DFC for his work as OC No 1 Squadron. Later in the year he was given command of a Spitfire Wing in Malta, but was killed soon afterwards while leading a sweep over Sicily. The Luftwaffe dropped a message over his airfield to say that he had been buried with full military honours.

The departing CO and his two lieutenants were the guests at a dance and cabaret held in their honour on 25 April. Their postings marked the end of an era for the Squadron, for the last of its peacetime regulars had gone.

On 8 April the Squadron had moved temporarily to Croydon. Another posting notice arrived by the end of the month, and Sqn Ldr Brooker moved No 1 up the road to Redhill on 2 May.

In addition to its daylight activities during the spring, No 1 had flown many hundreds of hours by night. In spite of careful training, the night patrols had so far yielded no results. Although the main blitz on London was waning with the shortening hours of darkness, the Germans were still sending occasional heavy night raids against the capital. Without the help of airborne radar, the Hurricanes had little hope of catching the night raiders under normal circumstances, but on the night of Saturday 10 May, the circumstances were abnormal — there was a full moon and the visibility was unlimited. And the Luftwaffe chose that night to send 550 sorties to London.

No 1 Squadron was brought to readiness at dusk as the coastal radar stations began plotting heavy concentrations of enemy aircraft forming up over France. Twelve of the Squadron's 15 Hurricanes were serviceable, with a 13th being prepared in the hangar. Brooker decided that they should patrol in sections of four aircraft with each Hurricane separated by 500ft for safety in the dark. As night fell the CO was warned by Sector HQ at Biggin Hill to prepare to scramble his first section.

The bell rang at 23.00hrs. The CO, Flt Lt Bill Sizer, Moses Demozay and Sgt Plasil took off and fanned out over central London at heights of between 16,000ft and 17,500ft. Brooker saw bombs exploding on the city and caught a glimpse of a Heinkel outlined against the explosions. He dived in behind it and opened fire. The Heinkel reared up, burst into flames and dived out of control into a fire that had just been started by one of its compatriots. Unfortunately Brooker had shot it down from dead astern, and his canopy became covered with his victim's oil. Although he attacked three more bombers during his sortie, he was unable to fire effectively as he could see nothing through his windscreen. Meanwhile, Moses had seen a Heinkel trapped in searchlight beams over the East India Docks, so closed in to 30yd behind and attacked from slightly below. The Heinkel rolled on to its back and dived, smoking to the ground. The Frenchman followed it down and watched it explode in the Lea Marshes.

The second section, led by Flt Lt Jackman, was scrambled at 00.15. Sgt Dygryn fired several bursts at a Heinkel which eventually dived into the countryside to the east of London. Sgt Kratkoruky, another Czech pilot, spotted one over the Thames at Canvey Island and chased it away to Southend. When he lost sight of it the bomber was losing height at 2,000ft with both engines smoking.

The 13th Hurricane came out of the hangar at 01.00hrs, and 15min later Fg Off Robinson took it up to patrol over Redhill at 13,000ft. Before long he spotted a bomber heading home some way above him and climbed to the attack. He saw his bullets strike the German's fuselage and silence the rear gunner. Driving the enemy aircraft to low level, Robinson lost sight of it against the dark ground: he claimed the Heinkel as probably being destroyed, but was only awarded a damage.

The third section was scrambled at 01.30 in

Sgt Dygryn fought under the nom-de-guerre of Ligotiky to protect his family in Czechoslovakia. On the night of 10 May 1941 he flew three times, and shot down a bomber on each sortie over London.
Squadron Archives

the charge of Sgt Novak, who chased a Junkers 88 which he met while climbing to his patrol line. He saw the results of a three-second burst of fire hit the rear fuselage before the bomber dived away into the darkness and was lost. Sgt Prihoda saw a pair of Heinkels over the West End and closed in to 100yd behind them; he was just about to open fire when all three aircraft were bracketed by heavy ack-ack and he was obliged to break off his attack.

Plt Off Behal crashed during this third patrol and was killed. He had been heard to call on the radio that he had been hit and was baling out, but on the following day his body was found in the wreckage of his Hurricane near Selsdon Park.

Sgt Dygryn had been refuelling when the third section had scrambled. Five minutes later he took off again and patrolled Kenley at 15,000ft. After some time he saw a Heinkel going south, then closed in from astern to point blank range before opening fire. After a few seconds, black smoke and flames belched out of the pilot's cockpit, and the bomber spun down out of control.

Flt Lt Jackman took off to patrol base with Plt Off Raymond at 02.25. After stooging around uneventfully at 15,000ft, Jackman dropped down to 13,000ft and immediately saw a Heinkel heading for London some distance ahead of him. He chased it for about three miles and opened fire at 250yd. The bomber went into a steep diving turn to head for home and Jackman followed it round and fired again. The rear gunner returned his fire but Jackman got in a good third burst at 9,000ft: the Heinkel lurched wildly and went into a vertical dive, still spitting fire from its rear gun. It was found strewn over the fields some distance south of Redhill. Raymond too got a Heinkel about 10 miles south of base.

Sgt Dygryn had not yet had his fill. As a result of a quick turn round he took off again at 03.15, and after 30min of patrolling to the east of Biggin Hill he caught sight of a twin-engined aircraft silhouetted against the pale eastern skyline. It looked like a Junkers 88, and was flying southeastwards at full speed. After a long chase Dygryn caught up, identified it near Hastings at 16,000ft, closed to his customary point blank range and fired all his ammunition into the Junkers in several long devastating bursts. Flames licked back along the fuselage and it dived into the cold waters of the Channel. Dygryn landed back at Redhill at 04.00.

Some days later the coastguards at Southend confirmed that Kratkoruky's damaged bomber had crashed into the sea off Shoeburyness. This brought the night's score up to eight destroyed and one damaged — a figure that was rarely equalled by the specialist night-fighter squadrons.

Throughout the following week the congratulations poured in. Air Marshal Sholto Douglas, now C-in-C Fighter Command, and Leigh-Mallory, AOC No 11 Group, both sent complimentary signals. On 17 May, Sqn Ldr Brooker was awarded the DFC as a result of the action; the next issue of Squadron Routine Orders contained a paragraph from the CO which pointed out that his medal belonged by right to every member of No 1 Squadron.

Sgt Dygryn, under his nom-de-guerre of Ligoticky, became a public hero, being whisked away to London to meet Dr Benes, the exiled Czech President. Dygryn, who also downed a Bf109 on 16 May, came away from the interview with the Czech War Cross and the Medal for Gallantry. Moses Demozay was decorated by de Gaulle for his part in the night's achievement, his Croix de Guerre now sporting an extra palm leaf, and he was awarded the French Medal of Liberation.

In the meantime the Squadron had returned to its normal tasks of offensive sweeps and convoy patrols. On 21 May it escorted 18 Blenheims to attack some oil installations at Bethune, with No 1 flying rear top cover to the formation while the close escort was provided by Nos 302 (Polish) and 258 (New Zealand) Squadrons. From the French coast to the target the raid was under continuous attack by small formations of Bf109s. Two of the Germans were shot down by Sgts Kuttelwascher and Kratkoruky; others were damaged by the CO, Plt Off Raymond, Fg Off Velebnovsky, Sgt Nassvetter and Sgt Dygryn. Fg Off Robinson was attacked over the Channel by two Bf109s and was never seen again.

On 25 May, Moses Demozay was in the news again when he destroyed a new model of Bf110 over the Goodwin lightship — and yet another palm leaf was added to his Croix de Guerre. His left breast now sported so much vegetation that his nickname was changed to 'Oasis'.

On 1 June No 1 Squadron was moved to Kenley with the minimum of equipment for two weeks stay, and its place at Redhill taken by No 258 (New Zealand) Squadron. The move was designed to give the Kiwis some experience of defending the vital Biggin Hill sector by night as further heavy raids were expected with the waxing of the full moon, but both squadrons returned to their original bases on 15 June with little to report.

In contrast, the pilots were involved in a tremendous dogfight with Bf109s in mid-Channel on the day after their return from Kenley. The 12 Hurricanes were detailed at 17.00hrs to take over escort duties for a number of Lysanders engaged on air-sea rescue work in the Channel. When over Tonbridge, Sqn Ldr Brooker was advised over the radio that bandits were approaching the Lysanders. The Squadron went to full boost and charged out across the Kent coast at 3,000ft, just in time to relieve the previous escort of Spitfires, which were getting low on fuel. They drove away the attacking Bf109s and then pounced on an enemy Heinkel 59 seaplane that was engaged on similar duties to the Lysanders. Flt Lt Colin Gray and Plt Off Allan downed the Heinkel whilst the rest of the squadron took on its escorting *Staffel* of Messerschmitts.

The fight lasted for over 20min, by which time Plt Off Kopecky, Sgt Novak and Sgt Prihoda had each destroyed a Bf109. On the debit side Sgt Nassvetter got separated from the rest of the Squadron and was jumped by a reinforcing group of enemy fighters. He crashed into the sea and, although rescued and taken to Dover hospital, died on the following day.

Lt Moses Demozay was posted to No 242 Squadron on 18 June and went on to cover himself in glory, finishing the war as the second highest-scoring French pilot, with 21 victories. He had planned to go into politics after the war, but France was robbed of an able patriot when he died in an air accident in December 1945. Colin Grey was also posted out shortly after this and went on to become one of the top scoring RAF pilots of the war, winning the DSO for his work as a wing leader in North Africa.

On 21 June the Squadron scored another notable success on a 'Circus' operation to Desvres airfield in Brittany. On this occasion six Blenheims were being escorted by Nos 1, 258 and 312 (Czech) Squadrons. Because of a navigational error, two further squadrons failed to make the rendezvous, so the 'Circus' set out without its top cover. Heavy flak was encountered over the French coast near Boulogne, but the target was attacked without interference from enemy fighters — the Bf109s were waiting until the formation was turning for home, and swept in to attack out of the sun. Sqn Ldr Brooker counted over 30 Messerschmitts coming in, some of which he identified as the new Bf109F. The CO attacked one of these which he saw shooting at another Hurricane, giving it a quick burst from about 100yd. The Bf109's undercarriage came down, black smoke poured out and the pilot baled out rather smartly.

Plt Off Marcinkus, the only Lithuanian pilot to fight with the RAF, was flying as wingman to Sgt Plasil. He attacked one of the Bf109Fs with two short bursts fired head on. His leader later confirmed that the Messerschmitt had broken into small pieces just after Marcinkus had flashed past. Immediately afterwards, Plasil attacked a formation of 12 enemy aircraft from the rear, firing a short burst at their 'tail-end Charlie' from 150yd. Big chunks flew off it and it went down in flames like a falling leaf.

The running battle continued as they crossed out over the Channel. Prihoda claimed two Bf109s as probably destroyed and 'Kiwi' Collins sent one down into water near Boulogne.

Plt Off Kopecky was crossing the coast when he attacked a Bf109 successfully at 12,500ft. He was watching it spin down towards the beach when he was bounced by two of its friends, who wounded him in the throttle hand and punctured the Hurricane's glycol tank. He ditched in the sea near Folkestone and was picked up by a rescue launch. He was highly incensed on his return to Redhill to learn that no one had confirmed his kill. The total score for the mission was four Bf109s destroyed, four probably destroyed and two damaged; Plt Off Maranz, an American pilot, failed to return. Three further Messerschmitts were destroyed by various members of the Squadron on similar missions before the end of the month.

On 2 July, No 1 was moved to Tangmere and charged with the defence of Portsmouth and Southampton by night. It arrived at the unit's ancestral home to find a very different airfield from the one its predecessors had left in September 1939 — where once had stood neat messes and flowerbeds now lay piles of rubble. Tangmere had been hit very badly during the previous summer and was only just beginning to recover; for some time, all personnel had to sleep out at a temporary camp at Goodwood.

During daylight the Squadron flew a number of convoy escorts and often deployed to Manston to stand by for top cover duties if required by a squadron of air-sea rescue Lysanders. By night the pilots embarked on a very intensive course of interception training. Throughout the late summer and autumn they flew from dusk until dawn on every night when the weather was suitable. In addition to radar and searchlight co-operation, a lot of time was spent developing new interception techniques in conjunction with Turbinlite-equipped Douglas Havoc aircraft of No 1455 Flight, twin-engined light bombers with a spritely performance, in whose nose had been mounted a powerful searchlight. This was intended to illuminate the enemy night bombers, and two Hurricanes would fly in formation with each Havoc in order to destroy any enemy machines encountered. The system was not a success and was soon dropped when specialist night-fighters entered service in greater numbers.

During August the Squadron's Hurricane Mk IIs were replaced by Mk IIcs, which were equipped with four 20mm Hispano cannon in place of the Browning machine-guns. The night training continued into November when No 1 was declared fully operational as a night-fighter unit, the only snag being that there was very little to fight, as German aircraft were now coming over in smaller groups and were much harder to intercept on this side of the Channel. Then someone hit on the idea of destroying the bombers as they left their home bases under cover of darkness, and the concept of the night intruder was born.

Sqn Ldr James Archibald Finlay Maclachan DFC and Bar joined No 1 Squadron (designate) on 1 November 1941. So far he had fought quite a war. After scoring eight kills during the Battle of Britain with Nos 73 and 145 Squadrons he had been posted to Malta. During a dogfight with a Bf109 over the island in February 1941 he was badly wounded in the left arm and although he baled out safely, his arm had to be amputated above the elbow. With characteristic courage 'Mac' fought his way back to flying: within 15 days of the amputation he had flown solo. He was sent to recuperate in East Africa, but during the six months that followed he notched up over 250 flying hours. On his return to England he went to the artificial limbs centre at Roehampton and persuaded them to build him an artificial arm that would fit the controls of a Hurricane. By October the arm was ready, and Mac won back his flying category.

No 1 Squadron was re-equipped with the cannon-firing Hurricane IIc in August 1941. These aircraft were used with devastating effect during the unit's night intruder campaign in 1942. *Aeroplane*

'One-armed Mac'. Sqn Ldr J. A. F. Maclachlan lost his arm in combat with a Bf109 over Malta. He designed an artificial arm to fit the controls of a Hurricane, and was promoted to command No 1 Squadron at the age of 21. He was an outstanding night intruder pilot, scoring five kills over Luftwaffe airfields in France. After leaving the Squadron he commanded a Mustang unit, but he was shot down in France on 29 June 1943 when leading a joint operation with No 1. *Squadron Archives*

On 4 November Mac took over from Sqn Ldr Brooker who was posted to the Far East in time to distinguish himself in command of No 22 Squadron in the defence of Singapore and the subsequent retreat through Java.

The training for night intruder operations continued through the winter months. A trial intruder sortie was flown on 16 December by Sgts Scott and Smith, who encountered heavy flak and searchlights over northern France, but no enemy aircraft. The winter weather then set in to prevent further intruder experiments.

On the morning of 12 February 1942 the Squadron was called to an immediate and unexpected state of readiness. There was a tremendous flap on at Group; a special fighter sweep was being organised because the German capital ships *Scharnhorst*, *Gneisenau* and *Prinz Eugen* had broken out of Brest under cover of bad weather. By the time their escape was spotted they were steaming flat out up the Channel close to the French coast for maximum cover from the Luftwaffe and coastal artillery.

Torpedo-bombers of the Fleet Air Arm and RAF were detailed to attack the battle cruisers and several squadrons of cannon-armed fighters were ordered to disrupt the large force of escorting destroyers and E-Boats. Six Hurricanes of 'A' Flight, led by Flt Lt Raymond, took off at 13.30hrs and made rendezvous over Hawkinge with the Spitfires of No 129 (Mysore) Squadron. They crossed the Channel under a 300ft cloudbase and spotted three German destroyers off the Belgian coast. The ships immediately opened up with heavy flak, but Raymond turned his Hurricanes straight into the attack. A level strafing run at 50ft was made by three pairs of aircraft in line abreast, each pair engaging one of the enemy vessels. Opening fire at long range, they raked the superstructure and turrets of the destroyers with 20mm cannon. Plt Off Kuttelwascher, in particular, noticed his shells exploding at the base of the bridge and causing considerable damage. The pilots closed to under 100yd with guns blazing and chose the very last moment to pull up and vanish into the clouds.

Sgt Blair, a South African, was lost together with Sgt Marcinkus from Lithuania on 12 February 1942. The Squadron was sent to attack the capital ships *Scharnhorst*, *Gneisenau* and *Prinz Eugen* during their dash up the Channel in bad weather after escaping from Brest. Blair was shot down and killed; Marcinkus was captured, but was shot after the Great Escape from Stalag Luft III.
Squadron Archives

White Section, which consisted of Sgt Blair and Plt Off Marcinkus, did not return to Tangmere. Raymond had noticed a great ball of fire out of the corner of his eye as he made his attack, and it was assumed that Sgt Blair, a South African, had been hit by his target's guns. Marcinkus had ditched in the Channel and was taken prisoner. After some time 'in the pen' in France, the Lithuanian was sent to Stalag Luft III where he took part in the Great Escape. He was recaptured a day later and was shot in cold blood by the Gestapo.

Two days after the shipping strike Flt Lt Raymond was posted to an OTU. Plt Off Karel Kuttelwascher, a very able and respected pilot, was given the rank of acting Flight Lieutenant and took over 'A' Flight in his place.

On 16 February, the prewar link between No 1 Squadron and the Borough of Brighton was formally re-established. The Mayor was entertained to lunch in the Officers' Mess and, after the meal, was flown over his city by Sqn Ldr Maclachlan. Later in the day the Mayor's secretary was taken for a ride in the Magister by Plt Off Sweeting. However, the aircraft stalled while turning in to land and both occupants were killed in the ensuing crash. Fortunately for the Squadron's good relations, the civil coroner found nobody to blame and remarked that it had been a 'pure accident'.

The night intruder operations proper opened, like nearly every milestone in RAF history, on 1 April. Sqn Ldr Maclachlan and Flt Lt Kuttelwascher took off from Tangmere at 22.15hrs and crossed the French coast near Le Havre. The CO returned to base at 03.00 with nothing to report, but 'Kut' came home with a big smile. He had orbited Evreux aerodrome for some time at 3,000ft without seeing any activity, so pressed on to Melun where he found the flarepath lit up. After a few minutes he saw a flare being fired and noticed a Junkers 88 on the runway in the process of taking off. He slotted in behind it, watched it climb into the circuit pattern with its navigation lights on, and as it turned downwind he closed to 100yd and raked it with cannon fire. There was a bright flash from the starboard wing and the Junkers 88 dived into the ground. Kut then saw another similar aircraft on the runway and damaged it with a strafing attack. At this stage the aerodrome defences woke up and started firing, so the Czech decided that it was time to leave, and vanished into the night.

On the following night the CO destroyed a train near Cherbourg, but bad weather over France then delayed further intruder sorties for two weeks. While waiting for the weather to clear, the Squadron continued with its night training and defensive patrols. During this period two Hurricanes were lost in night accidents; Sgt Bachurek suffered severe head injuries and Sgt Vlk was killed.

Mac and Kut flew their next sorties to France on 16 April. Once again the CO was out of luck but Kuttelwascher found two Dornier 217s at St André and shot one of them down. The wily old Czech was making it sound so easy that Maclachlan decided to let some of the other pilots have a go. On the nights of the 24th and 25th, Flt Lt Scott, Plt Off Perrin and Sgts Pearson and Cooper were sent out, but they all returned with nothing to report. Another demonstration by the two experts was obviously required.

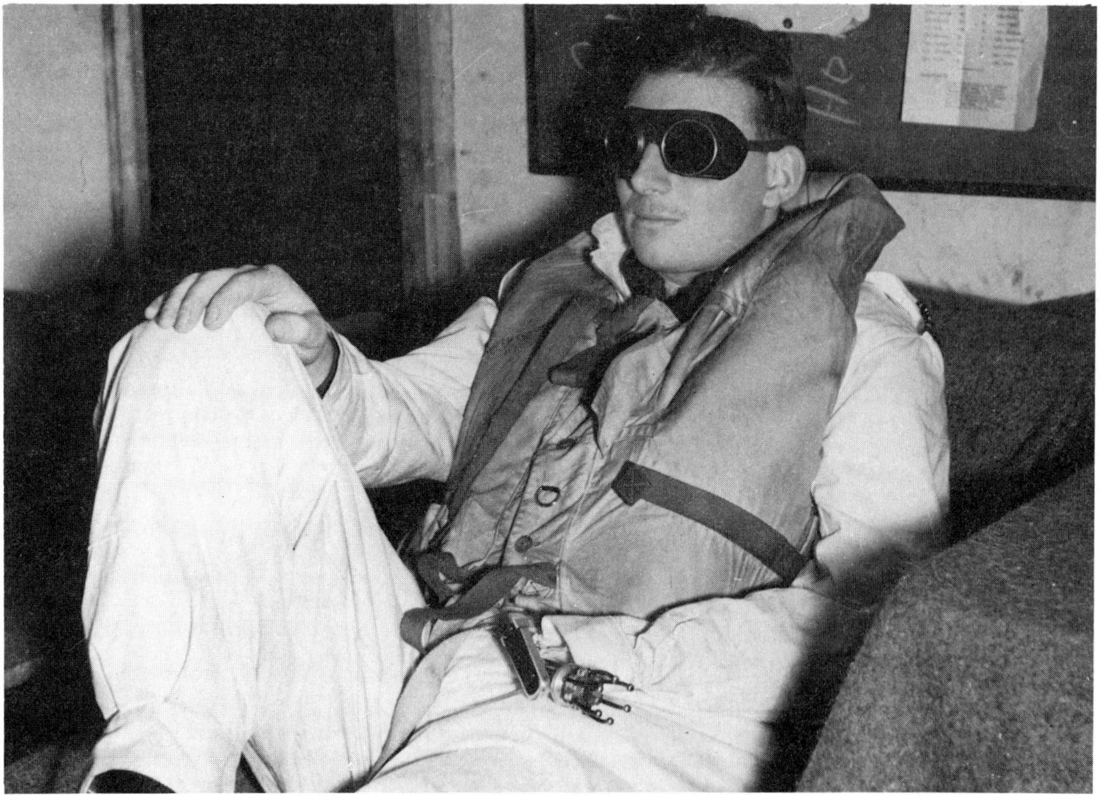

Sqn Ldr Maclachlan in night goggles, waiting to fly on a night intruder sortie. Good night vision was essential for intruder pilots, and very few people excelled in this role. *IWM*

On the following night the CO visited St André, destroying a Do217 and damaging another, while Kut went to Boos aerodrome near Rouen (from where No 1 had operated briefly in June 1940), shooting down a Dornier. He was just watching it crash among the hangars when red tracer fire flashed past his cockpit from behind. He throttled back sharply and dived, in time to see a Junkers 88 night fighter shoot past 30ft above him. He therefore jettisoned his long range tanks and pulled up on to the tail of the Junkers 88. After four short bursts of fire the German aircraft rolled to the left and dived away into the darkness. He did not see it crash, so only claimed it as being damaged.

Over the next few nights Sgt Pearson was sent out again, and three other pilots flew their first intruder sorties. They patrolled the same areas in which the CO and his 'A' Flight Commander had scored, but came home infuriated without having seen any enemy aircraft.

It was left to the Boss and Kut to finish the month in style. On the night of the 30th Mac saw no aircraft, but shot up two trains and a tug on the French coast. Meanwhile, Kut went further south than usual and destroyed a Dornier 217 which was taking off from Rennes. On the way home he looked in at the airfield at Dinard where two white lights were showing near the runway, and a few minutes later a Heinkel took off with its full navigation lights on. As Kut closed in the lights were switched off, but by then he had the Heinkel silhouetted against the moon, attacking it at 1,000ft with several long bursts of fire. A large piece of metal broke away from the starboard wing and the bomber dived into the sea with a picturesque moonlit splash.

May began with a series of bright and placid spring nights. Placid, that is, for all save the train drivers or northern France, who had a pretty thin time of it at the hands of No 1 Squadron's less experienced intruder pilots. The new boys were sent over to cut their teeth under relatively easy conditions, attacking 17 goods trains in Normandy and Brittany during

A 'night school' in No 1's crewroom at Tangmere in 1942. From left to right: Sgts Prideaux-Brune, Travis and Smith, Flt Lt Crabb, Plt Off Corbett and Sqn Ldr Maclachlan. *Squadron Archives*

the first three nights of the month; 15 of them were left with their engines belching steam or their trucks on fire. FS Bland in particular took to the sport of train-busting with enthusiasm. On 1 May he destroyed the engines of five goods trains before running out of ammo.

During these spring nights of bright moonlight, the CO was loath to send his more junior pilots out hunting over the heavily defended airfields. That task was left to the lot of himself, Kuttelwascher, Connolly, the two Scotts, Murray and Pearson. The others were promised a go as they gained experience, or when the nights grew darker with the waning moon.

On the night of 3 May five Hurricane IIcs crossed the Channel in perfect conditions: the operations of the previous week had obviously taken effect, for not a single train was to be seen. Murray and Flt Lt Scott orbited various blacked-out airfields but saw nothing. For Mac and Kut, however, the story was very different.

The CO found no activity at Rennes and flew up to Dinard, Here he ignored a dummy flarepath which had been laid and was just in time to clobber a Dornier that was preparing to land. There was no reaction from the defences and it would seem that they thought the Dornier's crash to be a landing accident. Maclachlan then noticed a Heinkel turning in with its wheels down, so he lined up for a quick burst. The nose of the bomber dipped, and it dived into the undershoot with one engine ablaze.

On the following night Karel Kuttelwascher was stooging around over Evreux and St Andre when three searchlights flicked up at him, but they were turned off when he flashed his nav-lights. At St André he came upon six He111s orbiting the well-lit aerodrome at 1,500ft. He joined the circuit behind the last one and shot it down. Moving into the resulting gap, Kut closed on the next Heinkel and sent it flaming into a wood to the east of the airfield. The remaining four Germans were still unaware of his presence, so he moved up one more place and destroyed a third aircraft. At this stage, the lights on the ground went off, a lot of light flak started coming up, and Kut beat a hasty retreat to Tangmere.

Flt Lt Karel Kuttelwascher was the most successful night intruder pilot of the war, scoring 15 kills over German airfields in France during his tour with No 1 Squadron in 1942. Both he and Maclachlan hunted alone. *R. Darlington*

The successes of Sqn Ldr Maclachlan and Flt Lt Kuttelwascher over the previous weeks had been widely reported in the British press. Some reporter with a good nose for an 'angle' had developed his daily reports around a supposed rivalry taking place between the two pilots for the highest score. 'Old Kuttel is three ahead now', 'Pilots duel for top score', 'One-armed Mac strikes again!', read the headlines. Maclachlan, a rather shy man socially, was quietly amused by all the nonsense, while the gruff Kuttelwascher just grunted at each successive news report, and shrugged it off. A more worthy acknowledgement of their work arrived on 16 May when the CO was awarded the DSO for his intruder operations and Kut got his first DFC.

Intruder operations continued sporadically throughout May, and by the arrival of the next full moon, all the pilots were fully experienced in the art. It had been found that pilots of equal day-time ability took to the intruder role with varying success. Night vision played a part in this, but the most significant factor was basic cunning, because for the only time in the Squadron's history, No 1 was doing a job that required no teamwork in the air. When skulking in the French darkness, the pilots were on their own, and their success and survival depended on hunting instinct and sixth sense.

On the moonlit nights of the first week in June the Hurricanes of No 1 Squadron ranged deep beyond the coastlines of occupied Europe. On the first night of the month the CO, Kut, FS English and Sgt Higham patrolled and strafed in their old hunting-grounds in Brittany. The two Scotts, Plt Off Perrin and Sgt Dennis went further afield, refuelling at Manston and trying their luck over Holland and Belgium. WO Scott damaged a Junkers 88 at Nieuport, while Perrin and Dennis claimed an E-Boat apiece near Ostend. Sgt Pearson also refuelled at Manston. He crossed the Dutch coast near Flushing and found a brightly lit airfield at Gilze where a bomber was coming in to land with its nav-lights on. Pearson could not identify it, but he shot it to pieces on its final approach to land.

This pattern of operation was repeated throughout the week. English destroyed a Dornier 217 at Eindhoven on the 2nd, but was killed in an accident over Bognor on the following night.

Mac and Kut both added to their score on 4 June. The CO got in amongst a group of orbiting Dorniers at St André, destroying two and damaging two more before his presence was noticed. Kut arrived on the scene an hour later as two Heinkels were landing and after shooting one down, turned his attention to some Dorniers which had appeared on the scene; he destroyed one and damaged another. As there was no flak, he next crossed the airfield at low level and shot up some firemen who were attending to one of Maclachlan's victims. The evening's work was spoilt by the loss of WO Dygryn, the Czech pilot who had just returned to the Squadron after a spell with No 210. Kuttelwascher was returning from his sortie and saw a fighter being shot down by flak at Le Havre at about the time when Dygryn was due to have crossed the French coast.

The success of the week's operations could be judged by the congratulatory signals that flooded in to the CO. Apparently, the Germans had diverted much of their night-fighter effort to cope with the intruding Hurricanes; the heavy Bomber Command raids on Essen and Cologne had got through with little trouble as a result. This was news to No 1 Squadron, as only one night-fighter had been seen, but the signalled thanks from 'Bomber' Harris were no less welcome as a result. From Sir Archibald Sinclair, the Secretary of State for Air, came the following message: 'Congratulations on your fine exploit last night'.

The Czechoslovak Inspectorate in London had been taking a keen interest in the deeds of No 1 since its countrymen had begun to fill the Squadron's ranks. A signal arrived on the 5th that read: 'Heartiest congratulations on your Squadron's last night's great successs. Good luck in future hunting to you and "Old Kut". AVM Janousek'.

'Old Kut' claimed a Ju88 at St André on the 22nd, and damaged another. The Germans at this particular airfield must have grown rather tired of his efforts by now; their sentiments were not shared by the King, who awarded Kuttelwascher a Bar to his DFC on 27 June. As Kut still had relations living in Czechoslovakia the award was not published until the end of the war, which seemed a little strange because his exploits were being reported daily in the British press.

On the night of 26 June, Plt Off Desmond Perrin was intruding over Holland. Finding no activity at Gilze or Eindhoven, he flew east in search of trouble. Again he found nothing, but in the course of his sortie he penetrated German air space as far as Dusseldorf before landing at Manston with dry tanks. This was the first time in two years that an aircraft of No 1 Squadron had flown over the soil of the German Fatherland.

Several new pilots had been posted to No 1 Squadron during its intruder phase, including Cdr André Jubelin of the Free French Navy. 'Ju', as he became known, had escaped from the Japanese in Indo-China in 1940 by flying a light aircraft to the Philippines.

Another round of successful intruder operations took place under the moon of early July. Sgts Campbell and Pearson each damaged a Dornier over Brittany, and Kuttelwascher was at it again; he destroyed two Dorniers at St Andre on the 2nd, and damaged a third. Kut's forays had been getting longer and longer as the months had passed. On one occasion he was airborne for over three hours — 10 minutes longer than the theoretical maximum endurance of the Hurricane with long range tanks. The other pilots used to pull his leg about this by claiming that Kut did not like landing at night, and was trying to stay airborne until dawn.

At the beginning of July it was learned that the Squadron was to be moved to Acklington in Northumberland to re-equip with the new Hawker Typhoon. The move would take place on 8 July, and would mark the end of No 1's night intruder operations. In the course of three months' hunting over France and the Low Countries, the pilots of No 1 Squadron had destroyed 22 enemy aircraft and damaged a further 13. Of those destroyed, five had fallen to the guns of the Squadron's young Commanding Officer, while no less than 15 had been shot down by Flt Lt Karel Kuttelwascher.

When No 1 left Tangmere, Kut left the Squadron. He later flew Mosquitos with No 23 and spent some time touring the United States giving lectures about the Royal Air Force. After the war he went home to Prague but returned to England when the Communists took over. He flew with BEA for many years, and died in 1959.

CHAPTER THIRTEEN

The Typhoon

The 16 Hurricanes of No 1 Squadron were led to Acklington on 8 July by Sqn Ldr Maclachlan. They arrived over the new airfield flat out at 50ft before pulling up to give the locals a display. Cdr Jubelin was flying as wingman to the CO. As the aircraft streamed in to land in pairs, the Frenchman tucked very closely in to his leader's side to make their arrival look smart. He overdid it a bit and landed with his wingtip astride Mac's rear fuselage: no serious damage resulted, and the incident went unnoticed by the spectators. The Squadron's first Typhoon arrived on 21 July and the re-equipment and conversion to the new fighter was completed by mid-September. The Hawker Typhoon was a formidable aircraft. Powered by a 2,000hp Napier Sabre engine, it could attain an indicated air speed of 340mph in level flight at 15,000ft and considerably more in a dive. It was very heavy, but it handled nicely both at speed and when landing, although the immense torque from the engine required careful handling when overshooting from a bad approach. No 1 Squadron's pilots were impressed with its speed but one remarked that it waddled like a duck in the circuit, and that the brakes were dreadful.

As with most new aircraft, there were teething troubles. A number of structural failures in other squadrons caused the Typhoons to be grounded on several occasions while essential modifications were carried out to prevent the tails falling off. The Squadron's Engineering Officer, Flt Lt 'Chunky' Chown,

A Hawker Typhoon IB at Acklington in late 1942. This eight-ton fighter was popular with its pilots, despite a poor rearward view and terrible brakes. It was used by No 1 Squadron against FW190s over the Channel and as a dive-bomber against V-1 sites in France. *Squadron Archives*

was sent away for courses with the manufacturer and returned with a lot of valuable 'gen' for the groundcrew. Chown had a reputation for being a maniacal slave driver, it was said that had the rest of the RAF worked like him, the war would have finished a year earlier. The NCOs in charge of his groundcrew were W/O Hicks and the very popular FS Sandy Sandham.

On 22 July Sqn Ldr Maclachlan was posted to No 59 OTU. The news came as a complete surprise to the Squadron, which had fondly anticipated taking the Typhoon to war under his leadership. Mac had earned the utmost respect and affection of all his men during the nine months in which he commanded No 1 Squadron, and André Jubelin later wrote of him that 'of all single-minded men Mac seemed to me the purest and loftiest I had ever met'. After a tour in the USA, Mac took over No 132 Squadron, flying Mustangs. He had relinquished his command of No 1 at the age of 22!

After a few weeks under the acting command of Flt Lt E. J. Harrington, the Squadron was taken over by Sqn Ldr R. C. Wilkinson. 'Wilkie' had joined up as a fitter in 1930, and as a Sergeant Pilot in France he had shot down nine German aircraft, winning two DFMs and a Croix de Guerre in the process. He came to No 1 after briefly commanding a unit of Hurribombers, when he had been shot down over Abbeville, and was awarded the OBE for his rapid escape from France. Long after his arrival as CO he was still receiving mail addressed to Sgt Wilkinson!

The two Flights were now commanded by Flt Lts L. S. B. Scott and F. W. Murray, who had both claimed several trains in the latter days of the night intruder operations.

By early September, the pilots and groundcrew were getting accustomed to the various foibles of the Typhoon. Fg Off 'Zulu' Watson from South Africa recalled that the early aircraft needed a modification to the sleeve valves of the engine, which were apt to fail. 'I was given explicit instructions by the CO as to what to do if the motor packed up. Sure enough after 5 hours on type my motor *did* pack up and I crash-landed near Charterhall — those instructions saved my life!'

Enemy activity was sporadic at the time and several scrambles had been ordered against fast 'tip-and-run' German raiders, but none had been caught. On 6 September Plt Offs Perrin and Bridges of 'B' Flight were sent off to investigate a radar plot approaching the Tyne. They climbed to 28,500ft — the absolute ceiling of the Typhoon — and spotted two

Three stalwarts of No 1. From left to right: WO Hicks and FS Sandy Sandham were in charge of the groundcrew from 1942 to 1944. Flt Sgt Mackintosh won the VC during World War 1, and was in charge of discipline on the Squadron. *C. H. Watson*

Sqn Ldr R. C. Wilkinson DFM and Bar commanded No 1 from July 1942 to May 1943. He had been promoted so fast that he was still getting mail addressed to Sgt Wilkinson long after he arrived. He led from the front and was greatly respected by his men. *C. H. Watson*

Flt Lt F. W. Murray, who commanded 'A' Flight at Acklington and Lympne, had also been a successful night intruder pilot at Tangmere under Maclachlan. *D. G. S. R. Cox*

fast-moving aircraft heading towards the coast near Blythe. They gave chase and identified their targets as two of the new Messerschmitt 210s. The Germans split up on their approach and the Typhoons attacked individually.

Bridges followed his target down the coast towards Whitby in a steep dive. He saw his cannon shells strike the enemy's fuselage, but then blacked out in a turn and spun down to 4,000ft. The 210 crashed at Robin Hood's Bay.

Perrin's Messerschmitt too had tried to escape southwards. He gave it a short burst at 25,000ft and saw pieces fly off its starboard engine. Its rudder caught fire and the enemy aircraft went into a vertical dive. Perrin followed down at high speed and blacked out when he tried to recover: he remembered that as he lost consciousness the needle on his ASI had been quivering around 520mph. Corrected for altitude and temperature this would have given Perrin a true air speed of well over 600mph. Small wonder that he had difficulty in pulling out — the thick wings of the Typhoon must have been covered in sonic shock waves. When he came round, the Typhoon was flying straight and level at 3,000ft. Several panels had torn away and his cockpit door had broken loose and was hammering wildly in the slipstream. As Perrin returned to Acklington he spotted a launch speeding towards the wreckage of the Messerschmitt he had destroyed.

These two kills brought the total number of enemy aircraft destroyed by No 1 Squadron to 230 since the outbreak of war. This made the Squadron the highest scoring unit in the RAF at that time.

Throughout the autumn and winter, the training programme continued at Acklington. Considering the many accidents suffered by other Typhoon units, No 1 reached operational status with a relatively trouble-free run. One fatal accident was a mid-air collision that occurred on 21 October, over the sea near Amble, when Plt Off P. N. Dobie and Mr Gerry Sayer (the famous Chief Test Pilot of the Gloster Aircraft Co) were carrying out air-to-ground firing experiments. Both men were lost without trace.

In the early days of February 1943 the news arrived that the Squadron was moving south once more, this time to Biggin Hill. The ground party moved off at 05.00hrs on 9 February, and ran into trouble at Newcastle. The RTO at the station was still half asleep and waved off the train before most of the NCOs had boarded. To make matters worse, several truckloads of ground equipment were diverted to Hull and Norwich.

The Typhoons also had their problems on the way down. Bad weather forced them to divert to Wittering, where Sgt Crowley ran out of hydraulic pressure and turned over on landing, fortunately without injury.

After arrival at Biggin Hill the pilots were sent out to learn the area. Sgt Bob Hornall had a lucky escape when he suffered engine failure over Sussex, because the countryside below was thickly wooded and full of tented army encampments. Hornall did extremely well to put his aircraft down in a tiny field without damage to himself or the nearby troops. The CO later received a letter of appreciation from the GOC 5th Canadian Armoured Division, who expressed his thanks for Sgt Hornall's skill.

Another tragic mid-air collision occurred on the evening of 6 March. Plt Off Slim Whitmore and Sgt Harris Fraser, an Australian, were returning from a patrol in close formation in thick cloud. Somewhere in the cloud they must have touched, for the wreckage of both Typhoons was discovered five miles apart, in the hop fields of Kent. Fraser's twin brother

Charles was also a pilot with No 1 Squadron at the time, so the accident brought a particular sense of personal loss to the Squadron. Both pilots were buried at Orpington on the 10th with full military honours.

No 1 Squadron remained at Biggin Hill for only seven weeks. The station was commanded at the time by Gp Capt Sailor Malan, who did not want Typhoons on his station and with whom Sqn Ldr Wilkinson did not get on. Wilkie therefore requested a further move, and was given the choice of Lympne or Gravesend. He chose Lympne because it was closer to the Germans and further from Malan! Not much else happened at Biggin Hill until the day before the move to Lympne.

On 13 March, Flt Lt Scott and Sgt Hornall were patrolling a line 10 miles south of Beachy Head. After 30 minutes, 'Cowherd' control at Kenley vectored them on to a pair of Focke Wulf 190 fighters in mid-Channel, with the results that can be seen in the accompanying combat report (overleaf).

The move to Lympne was completed by 15 March and the new station proved to be immensely popular with the Squadron personnel. From an operational point of view, they were right in the front line, down on the Kent coast between Folkestone and Dungeness — and they had also been deposited into the lap of luxury. The Squadron found itself accommodated in three vast mansions overlooking the sea. The grandest of all, the Officers' Mess, was Port Lympne, a beautiful house owned by Sir Philip Sassoon. Fountains played in the grounds, and life took on a style of unaccustomed elegance.

This cartoon of No 1 Squadron's badge was presented by Arthur Ferrier to the comedian Oliver Wakefield, who was the Squadron Intelligence Officer in 1943. *Frank Churchett*

SECRET

INTELLIGENCE FORM "F" DEFENSIVE PATROL (Type of Operation)

(This report is to be returned for every combat in which a claim has been made, and will also be used for offensive operations, if more information is available than has been previously returned in the composite report)

From: Sector Intelligence Officer, R.A.F. Biggin Hill, Kent.

To: Air Intelligence, H.Q. 11 Group and Air Intelligence Fighter Command.

STATISTICAL

Date:	(A)	13th March, 1943
Unit:	(B)	No. 1 Squadron
Type and Mark of our aircraft	(C)	Typhoon IB
Time attack was delivered	(D)	1725 hours
Place of attack and/or target	(E)	20 miles S.E. of Brighton.
Weather	(F)	Haze. About 5 miles visibility
Our casualties - aircraft	(G)	Nil
Our casualties - personnel	(H)	Nil
Enemy casualties in air combat	(I)	2 FW.190's Destroyed
Enemy casualties - ground or sea target	(K)	N/A

REPORT. Blue section took off at 17.05 hours to patrol South of Beachy Head under Kenley control. When about ten miles SW of Beachy they were given vectors on to E/A. These were seen at zero feet on the starboard beam, and the section turned towards them and came up on their tails. The E/A turned seawards and started climbing rapidly in echelon starboard. Blue 1 (F/Lt. Scott) took the right hand FW. 190 and Blue 2 (Sgt Hornall) the other as they climbed after them. Blue 1 fired a short burst at about 10000 feet from 250 yards range. The E/A appeared to explode in dense black smoke and pieces flew off. The E/A was next seen in a gentle dive to port with white smoke pouring out. Blue 1 had followed it and had another burst at 150 yards range. Strikes were seen, and more pieces, including the cockpit hood, flew off. The E/A was now gliding and employing clever evasive action, which spoilt Blue 1's third short burst, although more strikes were seen. Blue 1 then made a wide circuit and came in behind the E/A again, firing from 150 - 50 yards range. Strikes were seen all over the E/A which turned on it's back and crashed into the sea from about 2000 feet. While this combat was going on Blue 2 (Sgt Hornall) had opened fire at the other FW.190 at between 10000 and 13000 feet from 400 yards range. Smoke was seen to pour from the E/A and a piece fly off. The E/A then went into a shallow dive with smoke pouring from it. The FW.190 then went into a spin and Blue 2 gave it a second burst and it went spinning into the sea. Blue 2 then turned quickly and noticed Blue 1 making his attack. The E/A was emitting smoke and diving slowly. Blue 2 gave it a burst but saw no strikes and overshot. By the time he had turned both Blue 1 and the E/A were lost in the haze. Both FW.190's had a vertical red stripe on the fin and their undersurface was painted yellow.

Claims and Ammunition

		Cannon	Camera
F/Lt. Scott	1 Squadron	340	Not fitted
Sgt. Hornall	1 Squadron	493	Yes

Signed. J.M. Meadow F/O.
for Sector Intelligence Officer
R.A.F. Station, Biggin Hill

Sir Philip Sassoon's house, Port Lympne, which was the No 1 Squadron Officers' Mess from March 1943 to February 1944. *Squadron Archives*

The aerodrome had been owned before the war by Mr and Mrs Davis. Mr Davis had been killed shortly before the Squadron arrived, but his widow made sure that the pilots and groundcrew were well looked after at Lympne. She ran the NAAFI van around the dispersals, dispensing good cheer and excellent tea, and in the evenings her prewar liquor licence (which had been overlooked) came in handy. Zulu Watson remembered going to a party at her house and drinking whisky out of a small barrel.

The Squadron had not been forgotten by the citizens of Brighton, for Sqn Ldr Wilkinson and three of his pilots were invited to their 'home town' as guests of honour in the Wings for Victory celebrations, returning well-fed but minus the new Adjutant: Fg Off Tony Scrope-Davies had become carried away among the VIPs, and returned somewhat damaged the following day.

On 9 April White Section (Plt Off Higham and Sgt Sutherland) were scrambled from Lympne under Biggin Hill control to investigate low-level contacts near Gap Gris Nez. They saw four FW190s escorting a destroyer and roared in to attack them. Ten further Focke-Wulfs then materialised, so they took violent evasive action and got away. Black Section then arrived from Lympne and engaged five of the enemy at 200ft. Sgt Bob Hornall (Black 2) shot down one of these, escaping into low cloud before the others could catch him.

No 1 Squadron operated in this role for the following three months. Seemingly endless Channel patrols and standbys made the pilots somewhat envious of the Typhoon fighter-bomber squadrons that used Lympne as an advanced refuelling ground. As time went on, however, a number of missions were flown in escort of 'Bombphoon' squadrons over northern France. Plt Off Ernie Glover was shot down and captured on one such escort on 19 May, but the score was evened by Sgt Ramsey on the 23rd when he destroyed a FW190 at low level off Hastings with the help of Fg Off Zulu Watson.

When returning from escort sorties into France, the pilots took to train-strafing on the way home. Plt Off 'Moose' Mossip, a Canadian damaged two trains near Dunkirk on 5 May and left three barges in flames at Furnes. Early on 16 May, Flt Lt Scott and Plt Off Chalifour were sent over on a special train-busting sortie during which they ranged around the Douai area for over an hour without opposition and destroyed five railway engines in the process.

On 23 May, Sqn Ldr Wilkinson was promoted and posted to command the fighter wing at Gravesend, so Scott, as senior Flight Commander, assumed temporary command of

FS Crowley and FS Ramsay (standing) find something to laugh at. Ramsay had a lucky escape when he hit a telegraph pole while train-busting in France. *C. H. Watson*

Fg Off 'Moose' Mossip, from Canada, was an enthusiastic train-buster. His work won him a DFC in September 1943. *Squadron Archives*

No 1 Squadron, pending the arrival of the new CO. At the end of the month Sqn Ldr Tony Zweigbergk, a former Swissair pilot, arrived at Lympne to take up his command of No 1 Squadron.

More trains were attacked successfully by the Squadron's pilots during June. Mossip and Ramsey were well to the fore on several occasions, although Ramsey hit a telegraph pole during an attack on the 2nd: he had to fly back to Lympne in cloud with his engine cutting intermittently. Flt Lt Scott and Plt Off Dennis were not so lucky on the 16th. They had accounted for nine locomotives between them but were bounced by a group of Fw190s on the way home. In the course of the running battle from Bethune to Calais, Scott was wounded in the arm and Dennis was shot down and captured.

On 29 June the CO led four Typhoons to escort a pair of Mustangs that made a daring raid on a German flying school near Paris. The Mustangs were led by Sqn Ldr Maclachlan, now back from this tour of America. The Typhoons created a diversion at Creil while Mac and his wingman, Flt Lt Page, went in and destroyed six aircraft in 10 minutes over the training field. Moose Mossip, not to be outdone, broke away over the French coast when the escort was complete to destroy six locomotives in the St Omer area, which brought the total number of trains destroyed by No 1 Squadron to 92. When they got back to Lympne they were devastated to hear that Maclachlan had failed to return. It was later learned that he had crashed in a wood and had not survived.

On several nights during August, the CO, Mossip, Scott and Des Perrin flew up to Manston to join squadrons based there in their night intruder sorties, but on every occasion there was something amiss, with either the weather, the moon or the organisation. As a result the only intruder sorties carried out were those to the pubs of Hythe and Folkestone.

On one August morning, it was learned that a soldier had strayed into a minefield on the nearby Romney Marshes, and had been badly injured. Cpl Golding, the Squadron's medical orderly, dashed over in a jeep with two other men to try to rescue him and without thought for their own safety they gingerly penetrated the minefield and brought the soldier out on a stretcher. All three were awarded the George Medal for their courage. 'Doc' Golding already held the BEM and so for a while became the most heavily decorated member of the Squad-

The Adjutant and his 'gen-merchants'. Fg Off Tony Scrope-Davies poses with his admin team outside the orderly room at Lympne. On his left is FS Mackintosh VC, and beyond him, Cpl 'Doc' Golding BEM, who was later awarded the George Medal for his rescue of a wounded man from a minefield.
Squadron Archives

ron. That is of course with the exception of FS Macintosh, who had won the Victoria Cross during the Great War. In his service with No 1 he played the dual role of disciplinary NCO and Sqn Ldr Zweigbergk's cribbage partner.

Meanwhile, Fg Offs Mossip and Perrin and Flt Lt Scott had been awarded the DFC for their work in recent escort and train-busting operations. When Scotty was posted to another Typhoon unit on 23 August, Des Perrin was promoted to take his place in command of 'B' Flight.

Lympne was a tricky aerodrome from which to operate the heavy Typhoon. It was a rather small grass airfield, and as a result there were numerous incidents on take-off and landing. Zulu Watson recalled several:

'Back on 18 March, I returned from ops with no brakes and not much petrol. I overshot Lympne and dropped 5ft into a road and turned over in a field. I remember a WAAF running up to me and trying to stand on her head to look at me, hanging unhurt upside down.

'On another occasion I came back from ops, flying No 2 to the CO. My undercarriage lights were not indicating, so I went round again to land. I was not to know, and could not see, that the last bloke to land was taxying back up the middle of the airfield — we removed each others mainplanes. He left the Squadron shortly afterwards!

'I also remember seeing a replacement Typhoon being delivered to Lympne. It made a first-class landing, but as it taxied in we could see no pilot. In the end a slip of a girl got out, and she had blocks attached to her feet, so that she could reach the pedals — Diana Barnato!'

On 21 October the Squadron's first kill for some time was recorded. This was a Focke Wulf that was blown apart in a concerted attack by a pair of Canadian pilots, Fg Off Wiley and Sgt Fairbairn.

By the end of 1943 the Typhoons had been fitted with bomb racks and began to prepare for a more offensive role across the channel. The dive-bombing operations on which the Squadron now embarked were code-named 'Ramrods'. Two or three Typhoon squadrons would cross into France at low level with 500lb bombs tucked under each wing, and then climb to around 10,000ft before delivering their attack. The bombs would be dropped in a 60° dive from 5,000ft and the aircraft would join up on the deck to hedgehop their way home.

Experience had proved that these dive-bombing attacks could be executed with pinpoint accuracy. The 'Bombphoon' squadrons were accordingly tasked to attack point targets such as airfield hangars and some mysterious construction sites which had recently been observed in Normandy and the Pas-de-Calais — they were later identified as launching sites for the V-1 flying bombs.

The first effective 'Ramrod' was flown by the Squadron on 10 November. A Herculean effort by the servicing echelon produced 18 of the 20 Typhoons fully serviceable. The pilots of No 1 crossed the Channel at 50ft in company with No 609, before pulling up to 11,000ft to bomb a construction site at Martinvaast in Belgium. Most of the pilots were bombing for the first time, and found it very unnerving to hold their Typhoons in a roaring dive for almost 20 seconds as the flak burst all round them. Those who attacked last were gratified to see the bombs of their predecessors bursting all over the target. Mossip and Campbell were slightly damaged in the dive, but took their revenge by strafing troops on the coast on the way home.

'Ramrod' sorties of this nature continued until early February, when the Squadron was ordered to move to Martlesham Heath near Ipswich: by this time a total of over 400

Sqn Ldr A. Zweigbergk with his pilots at Lympne in late summer 1943. Flt Lt Des Perrin, OC 'B' Flight, is on his left, and the tireless Engineering Officer, Flt Lt Chunky Chown, is on his right. *C. H. Watson*

bombing sorties had been flown by the Typhoons. Few enemy fighters had been seen and none of these had ventured to attack. Most of their mysterious targets had been hit extremely hard and accurately by a total of some 200 tons of high explosive, but to offset this, four pilots had failed to return, although one of these, WO F. J. Wyatt, was later reported to be a prisoner of war.

The move to Martlesham was completed on 15 February. Every member of the Squadron was sorry to leave Lympne, because although within sight of the German-occupied French coast, life on the little airfield had been tinged with a curious holiday atmosphere. Their accommodation had been exceptionally luxurious, and many good friends had been made in the area — the airmen in particular were sorry to be taking their leave of Mrs Davis and her NAAFI wagon. When No 1 left Lympne the groundcrew contributed to an engraved silver cigarette case which they presented to their favourite 'Mrs Mobile'.

The new airfield was the home of a USAAF P-47 Thunderbolt Wing. The hectic activity and the American sense of high drama were very new to most members of No 1 but a lot of new friends were made and the Squadron had its first taste of an American PX. Sgt Algy Gower recalled that this emporium was as lavish as a West End nightclub, and Camels and Lucky Strikes cost 3d for a pack of 20. The NCOs and airmen shared 'chow' times with the Yanks, and gaped in disbelief at the sight of four to six eggs on a plate for breakfast.

During February and March the Squadron flew as close escort on 11 daylight raids by American bombers, shepherding up to 110 Fortresses, Marauders, Bostons and Mitchells to their targets. The Focke-Wulfs were frequently seen on these missions but they invariably made off when engaged by the escort, and not a single bomber was shot down by enemy fighters when the Typhoons were around. One of the Squadron's aircraft was lost on 2 March, when WO Neil Howard suffered engine failure near the target. He crash-landed successfully in France and was taken prisoner, subsequently escaping and returning to his unit a year later.

'Moose' Mossip had been posted to a Spitfire unit in February, but returned on 21 March to visit his former squadron. The CO let him borrow his old Typhoon for a nostalgic trip, but the Moose burst a tyre on take-off and completely wrecked the aircraft. He went back to his Spitfire feeling distinctly unpopular.

At the end of the month Sqn Ldr Zweigbergk, Flt Lt Higham and Fg Off Hornall were posted suddenly. The loss of three such experienced pilots came as a distinct shock to No 1 and foretold of an impending move. Flt Lt Des Perrin took acting command of the Squadron and learned a few days later that they were to go to North Weald on 2 April, to begin re-equipment with the Spitfire Mk IX.

CHAPTER FOURTEEN

The Spitfire

The remaining pilots of No 1 Squadron flew their 17 Typhoons into North Weald on the afternoon of 3 April 1944. Waiting for them to arrive were the new CO, Sqn Ldr John Checketts, DSO, DFC, and his 'A' Flight Commander, Flt Lt I. P. Maskell. On the day of the Squadron's arrival, four more Typhoon pilots were posted to other units to be followed shortly afterwards by the departure of the indefatigable Flt Lt Chunky Chown. Chunky had served for 2½ years with No 1 and his loss was felt very deeply by pilots and groundcrew alike.

No 1 remained at North Weald for three weeks. In that time, more Typhoon pilots left the unit, their places taken by Spitfire drivers of varying experience. Among these were four Danes: Flt Lt Pederson, Fg Off Christiansen, Sgt Jacobsen and Sgt Petersen. Inevitably perhaps, they became known as Petes I and II, Chris and Jake.

One of the last Typhoon boys to leave was Flt Lt Des Perrin, who had been with No 1 since the final days of the Hurricane. This immensely popular pilot was promoted on his appointment to the staff of a flying school at Annan, but lost his life later in the year when leading No 198 Squadron on a sweep.

For the first week, the new arrivals had to content themselves with flying a number of Spitfire Mk Vs, but 14 Mk IX LFs were delivered on the afternoon of 12 April, much to the delight of the pilots. The lads in the servicing echelon, still under FS Murphy, were not quite so ecstatic. To them fell the long hours of acceptance checks, painting the Squadron's code letters JX on each fuselage and readying the new machines for flying the next day.

Flt Lt Ian Maskell, OC 'A' Flight from April to December 1944. *D. G. S. R. Cox*

Sqn Ldr Henry Lardner-Burke with his pilots at Predannack, from where No 1 Squadron covered the D-Day landings. The group includes three Australians, three New Zealanders and two South Africans. *Squadron Archives*

For some time past a small group of the officers and men of No 1 had been building quite a reputation for themselves on the stage. This had started back at Acklington when the Squadron had put on a variety show called 'It's in the Air' to entertain the station at Christmas time. The tradition had continued at Lympne, where the small group of enthusiasts was led by LACs Tom Merry and Frank Churchett. They renamed themselves 'The Erkadians' and provided a lot of fun and laughter for the rest of the Squadron and audiences at the many stations they visited. The star of the show was a young airman named Max Bygraves, whose voice and humour gave the cast a distinctly professional air. The group remained almost intact through the moves to North Weald, where they made an immediate hit. Such was their popularity that the CO was persuaded to let them remain in Essex for a week after the Squadron moved on, in order to complete their full programme of engagements. 'The Erkadians' finished up with a cast of 40 and a three-hour variety show, and Max Bygraves won the Hayes Challenge Cup for the best entertainment act in Fighter Command.

By 20 April the Squadron was fully converted to the Mk IX and was ready to move up to Ayr for an armament practice camp. On leaving North Weald Fg Off Red Ward was killed in a mid-air collision with Fg Off 'Pussy' Cattermoul. Cattermoul's aircraft was badly damaged but he managed to bring it back to a safe landing at North Weald. The remaining 17 Spitfires and two Oxford trainers were delayed en route by bad weather at Church Fenton for two days, but the Squadron eventually managed to complete its course of air-to-air and air-to-ground firing by the end of the month.

Whilst at Ayr, two startling bits of news arrived. The first was that No 1, instead of returning to North Weald, was to proceed to Predannack in Cornwall to patrol against the Junkers 88s which were attacking shipping in the Western Approaches. The second snippet was that the CO was to be promoted, and posted away to command a Mustang Wing. Sqn Ldr Checketts had only been with the Squadron for four weeks but in that time had become a highly respected CO.

The move to Predannack took place on 29 April, with the Spitfires calling in at Rednal to refuel. At the Cornish airfield they were greeted by their new South African CO, Sqn Ldr Henry Lardner-Burke, DFC, who had served in Malta until 1942 and gained a personal score of 7½ kills.

At Predannack No 1, together with No 65 formed a Spitfire Wing under Wg Cdr D. G. Smallwood. For the following six weeks they ranged over the Western Approaches to the English Channel and the Irish Sea, in the vain hope of catching the marauding 88s. They were frequently scrambled to be vectored on to a convoy that was being attacked, but always arrived too late. These 'Instep' patrols and scrambles were interspersed with shipping reconnaissance flights off the French coast between Lorient and Cherbourg, accomplished with the aid of 90gal overload tanks.

On 19 May Fg Off Neville Brown led a section of four aircraft to reconnoitre Brest harbour. Seeing several heavy merchantmen and four destroyers moving out of the port, he requested a shipping strike. By early evening 'Splinters' Smallwood was airborne at the head of both squadrons as escort to a wing of Coastal Command Beaufighters. The convoy was spotted four miles off Ushant and was attacked with rockets by the Beaufighters. The escort, flying at 500ft, drew a curtain of light flak, but none of the Spitfires was hit. Two of the Beaus were not so lucky: one was shot down over the target and the other had to ditch on the way home. The strike sank one destroyer and damaged a couple of the merchant ships.

Three days later two of the Squadron's pilots were lost during a fighter sweep over Brittany. Flt Lt Alec Vale, who had recently taken command of 'B' Flight, was leading a section of four Spitfires which attacked a troop train near Vannes. During their second strafing run both Vale and Sgt Jacobsen were hit by small arms fire and shot down. Vale had been due to travel to London that night to receive his DFC from the King.

At dawn on 4 June, No 1 Squadron took off to give cover to a convoy assembling near Falmouth. On arriving overhead the pilots were amazed at the size of the gathering armada of light and heavy cruisers, destroyers, MTBs, corvettes and cargo ships which was assembling in formation off the Cornish coast. The promised invasion of France was obviously at hand.

The growing convoy was shadowed continuously by the Squadron's Spitfires for the next two days. On the afternoon of the 5th, all personnel were confined to the airfield and the returning patrols of the evening reported that the fleet had begun to move out into the open Channel.

By dawn on D-Day, Sqn Ldr Lardner-Burke was leading eight of his aircraft towards the French coast at low level. On the way they passed over a continuous stream of shipping loaded down with men, munitions and supplies, before swinging away to the south for an uneventful sweep over Brittany. After several convoy patrols during the day, they returned to Brittany at 15.00hr. Fg Off Wyllie attacked and destroyed a train near Kerlin Bastard and then saw a camouflaged staff car speeding up a country road. He made a frontal strafing run on the car and saw his cannon shells punching through the bonnet and windscreen. It burst into flames and dived into a ditch. Fg Off Stuart was hit by rifle fire but returned safely to base.

Allied fighter operations were severely hampered by low cloud throughout the following week. The pilots at Predannack sat and chafed, anxious to get stuck in over the beach-head, but they realised that little could be done to support the troops who were clawing their way inland. By 12 June the weather had cleared sufficiently to permit further operations, and several road convoys were strafed in Brittany.

On the 14th the 24 Spitfires of the Predannack Wing penetrated France as far as Le Mans. Fighter opposition had been expected, but every airfield they passed had been heavily damaged in recent bombing raids, and the Luftwaffe was notably absent. On the way home, Fg Offs Stuart and Marsh ran very short of fuel. Calculating that they would not reach their home airfield, they left the formation and refuelled at the first emergency strip on the beach-head near Grandcamp. They were among the very first fighters to do so, thus making the first landings by No 1 Squadron in France for nearly four years.

Over the next few days the Spitfires were armed up with 500lb bombs and sent to attack radar stations and shipping installations on the French coast. On 15 June they struck at the jetties of St Peter Port in Guernsey, where some U-Boats were reported to be sheltering. The bombing results were excellent. Severe damage was done to the jetties and coastal shipping, while the bombs fell clear of adjacent houses in the only part of Britain to be occupied by the Germans.

On the 19th the Squadron was moved to Harrowbeer, near Plymouth, to provide close support for the armies in Normandy. The Harrowbeer Wing was led by Wg Cdr H. A. C. Bird-Wilson, but he was not destined to see No 1 go into action from his airfield. Shortly before the Squadron had moved from Predannack, the first V-1 flying bombs began falling on southeast England. Various squadrons of high speed fighters were alerted to combat this new menace, and No 1 got its marching orders to Detling within 36 hours of arriving at Harrowbeer.

The Squadron's first anti-V-1 patrols — codenamed 'Diver' — took place on 24 June. The pilots were very keen to get their teeth into some explodable opposition, but the first few days were fairly chaotic as they tried to learn this new game. Several 'Doodlebugs' were seen, but the Spitfire pilots all had trouble in

Fg Off W. J. Bachelor, who destroyed the second of 46 1/6 V-1 flying bombs that were brought down by No 1's pilots. *D. G. S. R. Cox*

Sqn Ldr David Cox DFC and Bar commanded No 1 from 11 January to 21 April 1945. He had previously destroyed seven enemy aircraft during the Battle of Britain and in Tunisia. Winston Churchill asked specifically for No 1 Squadron, led by Cox, to escort him to Venlo to view the crossing of the Rhine. *D. G. S. R. Cox*

turning in on their targets, which were moving much faster than anything they had fired at in the past. To make matters worse, some of their attacks were baulked by Tempest pilots who had been in the fray for all of a week and who claimed priority for their faster aircraft.

No 1's patience was rewarded on the 27th by the destruction of two flying bombs. The first was chased for many miles and shot down near Wadhurst by Fg Off Bob Bridgeman. It had been flying at 2,000ft at an indicated speed of 375mph. The second bomb was downed in the afternoon by Fg Off Batchelor. He saw it crossing the coast over Hastings and waited until it was clear of the town before making his attack. It rolled on to its back and crashed with a tremendous explosion near Rye.

On the following day five more V-1s were destroyed by No 1's pilots. One of these kills must go down as the shortest successful fighter sortie in history. Fg Off McIntosh was scrambled on a 'Diver' patrol at 17.35hrs. He was just raising his undercarriage when he saw a flying bomb coming across the coast at low level. He gathered speed and turned downwind in a classic curve of pursuit on to his target. As the V-1 crossed his nose, he gave it one short burst using four rings of deflection — a massive crossing angle. To his delight the 'Diver' exploded in a great fireball. MacIntosh continued his turn, lowered his wheels and landed — at 17.36hrs!

Seven V-1s were destroyed on 30 June, two of them by Fg Off Ken Foskett during the same sortie. Sgt Hastings had a narrow escape when he pursued a flying bomb into the London balloon barrage. He had just destroyed his target in bad visibility when the balloons started looming up out of the murk. He pulled up to avoid them but struck one of the cables, which snapped off two of his propeller blades and sheared 6in from his starboard wingtip. He was very lucky to be able to make a successful forced landing at Gatwick.

After an absence of five months, No 1 Squadron returned to Lympne on 12 July. The pilots found their unit's former station a little crowded on their arrival: originally a single squadron station, Lympne now had to cope with almost 50 Spitfires of Nos 1, 165 and 41 Squadrons. Some jockeying for the best positions took place, but No 1 eventually secured its former dispersal in time for the afternoon visit of Mrs Davis and her NAAFI wagon, who was delighted by the return of her favourite squadron.

On 16 July, a complete reorganisation of the anti-doodlebug defences was brought into effect in which the fighters of the Air Defence of Great Britain, as Fighter Command had been renamed, were given patrol lines in mid-Channel and across the Weald of Kent. The ack-ack belt was moved to the South Coast, which was declared a prohibited area to give the gunners a chance to fire at anything that crossed their positions. This new plan was not at all to the liking of most fighter pilots. They found that they were chasing the 'Divers' into the gun belt and having to break away before getting the opportunity of a kill. In the early days of the new scheme, few bombs were brought down by the guns, and the percentage of those falling on London rose sharply until the experience of the gunners increased.

By 23 July Fg Off Davy had become the Squadron's top-scoring pilot when he destroyed four 'Divers' within 12 hours, to bring his total score to five. On the following day Ken Foskett brought down his third, on a railway line near Ashford. On turning back, he noticed to his consternation that a train was coming round the curve towards the ensuing crater. Foskett did everything humanly possible, short of landing on the track, to warn the train driver — he flew alongside the cab waggling his wings and lowering his undercarriage. At the last moment he was delighted to see the train pull up just short of the broken rails. Messages of appreciation from the passengers and the Southern Railway poured in, and Foskett became a national hero in the press on the following day.

On 10 August the Squadron received orders to return to Detling to resume its offensive role. A total of 46$^1/_6$ V-1s had been destroyed by No 1 during the previous six weeks.

For the remaining nine months of the war the Squadron was employed on a variety of escort missions. Fighter sweeps and armed reconnaissance were thrown to leaven the mixture, but most of the flying hours were spent in relative boredom, droning across western Europe at the wingtips of the Lancasters and Halifaxes.

The Squadron, in company with two others, flew its first heavy bomber escort on 11 August, when 120 Lancasters bombed the marshalling yards at Lens. Moderate flak was encountered over the target but no enemy fighters were seen.

The lack of opposition had created an air of complacency, for which No 1 was to pay dearly a few days later. While escorting 36 Mitchells to attack a fuel dump in France, the Spitfires were taken by surprise by four audacious Focke-Wulf pilots. The Germans dived in from the rear, hit two of the fighters and escaped before the Squadron realised what was going on. WO Wallace-Wells was able to limp his Spitfire as far as the Channel, where he ditched and was soon rescued, but WO McKenzie was not so fortunate. He crash-landed in the target area, but was able to radio to his leader that he had survived the impact. He lay doggo for a week and was rescued by advancing Allied troops before rejoining the Squadron in September.

In early September the Squadron was employed on armed reconnaissance in southern Holland. On the 10th it completely destroyed a large goods train at Uitgeest, while on the 15th it attacked a mysterious storage area to the east of the Hague. The site had been given the codename 'Big Ben' and was strafed for 10 minutes by the 12 Spitfires that took part. The target was heavily wooded and, although many vehicle tracks were in evidence, there was no activity to be seen until a huge explosion ripped the wood apart. The pilots did not know it at the time, but No 1 had become the first fighter squadron to destroy a V-2 rocket site. Just one week beforehand, the first of the big rockets had fallen on Chiswick, and had caused widespread devastation.

In the early hours of 17 September a special operation order arrived. The signal itself was 12ft long and contained the details of the greatest airborne assault of all — Operation 'Market', at Arnhem. For the nine days that followed, No 1 Squadron was heavily engaged in escorting some of the 4,000 aircraft and 500 gliders which ferried the 'paras' to their bloody destination. No enemy fighters were seen by the Spitfire pilots, but they were able to contribute by attacking a number of the flak positions that had taken a heavy toll of the transport aircraft.

Just before Christmas the Squadron moved to Manston. The station was already desperately overcrowded as the home of three other fighter squadrons and a number of lodger units. The only accommodation available was cramped and uncomfortable in the extreme — a far cry from No 1's pleasant surroundings at Detling and Lympne. This last wartime Christmas was made still more dismal by a spell of quite abominable weather. No ops were flown for many days, and the Spits dripped forlornly on the streaming tarmac.

On New Year's Eve it was announced that Sqn Ldr Lardner-Burke was to be promoted to take over the Coltishall Wing, and that he had been awarded a Bar to his DFC. Flt Lt Ian Maskell, who had commanded 'A' Flight since

April, was also promoted to take over No 91 Squadron, which was stationed across the airfield.

As another rainy dawn ushered in 1945, Sqn Ldr D. G. S. R. Cox, DFC and Bar assumed command of No 1 Squadron. David Cox had originally been posted in to take the place of Maskell, but he was given instant and popular promotion when his predecessor was posted out. The heavy clouds and snow of January prevented the execution of any bomber escorts, but the Squadron was tasked for a number of sweeps and armed recces in Holland when the weather permitted. On some of these missions several pilots spotted the dramatic vertical contrails of V-2 rockets lancing into the clouds, but apart from reporting the suspected launching sites, little of value resulted from their trips.

The weather improved sufficienty during February to permit an increase in offensive operations. The advance of the Allied armies towards the Rhine now permitted the Squadron to refuel at Ursel, in Belgium, before taking off for an op. Several attacks were made in this way on V-2 sites in the Haagsche Bosch near the Hague. These strafing and bombing sorties were invariably enjoyed by the participating pilots, but the frequent bomber escorts into the German industrial heartland were not. The Spitfires would pick up the heavies over Belgium and shepherd them at 30,000ft to the Ruhr, and once as far as Paderborn. The flak was always heavy, but few fighters were seen and the four-hour sorties were cold and boring. Once, over Mannheim, a small formation of the new Me262 jet fighters came whistling through at very high speed, but without bringing down any of the bombers. Although they were seen in time, the Spits had no hope of attacking them.

No 1 was also selected for special VIP escort duty and on three separate occasions had the distinction of providing close escort for the Prime Minister. When Mr Churchill travelled to Volkel in Belgium on 23 March he specifically asked that the Squadron be tasked with looking after him on the way. A squadron of American P-47s got very nosy on the way over and David Cox had to weave his Spitfires close to the PM's Skymaster to fend them off. After Churchill had landed, Cox brought No 1 past in close formation before landing at Ursel for the night.

On the following morning they realised the significance of Churchill's visit. They were ordered into the air at first light to patrol the length of a stream of 1,500 glider combinations making the historic assault on the Rhine. No fighters came up to challenge this airborne armada, but several of the tugs were hit by flak.

Sqn Ldr David Cox and Fg Off Neville Brown after an op. *D. G. S. R. Cox*

By now the end of the Third Reich was imminent. The strategic bombers which had helped bring the Nazi regime to its knees ranged unopposed over the Fatherland, escorted by clouds of unemployed fighters. On 18 April No 1 rode shotgun to 900 Lancasters and Halifaxes as they pounded the fortress of Heligoland. And then, on the 25th, came their last wartime operation in Europe. At 15.50hrs 11 Spitfire IXs took off from an advanced airstrip in Holland to escort 446 Lancasters to attack fortifications on the island of Wangerooge in the Friesians. The Lancs approached in great gaggles and bombed with deadly accuracy from 14,000ft.

And so it came about that the last tally-ho of war echoed down the clouded European skies. The news of the German surrender was greeted at Coltishall on 8 May with quiet rejoicing. The pilots and groundcrew who had striven so long for this day could hardly believe that it had arrived.

Suddenly, and almost as an afterthought, they were tasked four days later with a final operation — the liberation of the Channel Islands. As the naval landing force accepted the last German surrender, 12 Spitfires sang their song overhead. No 1 Squadron, the first fighter unit to fly away to the war, ended it with a flourish.

CHAPTER FIFTEEN

Blue Note

The first Spitfire F Mk 21 for No 1 Squadron was delivered two days before the war ended in Europe. This, the most graceful Spit of them all, arrived too late to see active service. It took several months to re-equip the Squadron completely, and the last Mk IX LF was not withdrawn until the following year.

On 14 May the Squadron was moved to Ludham, where two pleasant months of summer were spent in getting accustomed to the strange leisures of peacetime. Then it was time to move on again to Hutton Cranswick in Yorkshire.

Victory in Europe had been celebrated surprisingly quietly by No 1. Not so VJ-Day. All work was cancelled for two days to allow the boys to go out on the biggest binge of their lives, but they returned to the camp on the first night very disappointed, having drunk every pub they visited completely dry. This was not so great an achievement as it sounds, as there was a great shortage of beer. They made up for it on the following night at the Station dance, where the NAAFI revealed its hoarded war reserve now that peace had broken out.

Operated by the Squadron for the following few months were four special Spitfire 21s, equipped with twin three-bladed contra-rotating propellers. They were marginally faster than the standard 21 and had a better rate of climb. But the days of the piston-engine fighter were numbered in RAF front line service and this variant did not see large scale production. One of the quartet, JX:1, was written off in a landing accident in April 1946.

Sqn Ldr Nash was posted to the Middle East in February 1946 and handed the command of No 1 to Sqn Ldr H. R. Allen DFC. The two Flights of the Squadron were now led by Flt Lts W. J. Batchelor and R. Hesselyn MBE, DFC, DFM and Bar. During the same month, a party was held to celebrate the reunion of No 1

Two of the Squadron's Spitfire 21s over the Solent in 1945. The first Mk 21 was delivered just before VE Day, and the type did not see active service. The second aircraft, JX:U, is preserved today in the squadron museum at Wittering. *Squadron Archives*

Spitfire 21s drawn up tail-to-tail at Tangmere in the summer of 1946. Sqn Ldr Dizzy Allen is at the extreme left, contemplating the dogs.
Squadron Archives

Squadron with No 6001 Servicing Echelon, following a nominal separation of three years.

At the end of April, No 1 was finally allowed to come home to Tangmere. So much had changed since Bull Halahan had taken his Hurricanes to France, yet at heart it was the same old Tangmere and the same old No 1. Within days of the Squadron's arrival, a great winged '1' had appeared on the wall of a new hangar, which now stood on the site of the old one which had been bombed in August 1940.

The Squadron flew up to Acklington on 17 June for its first peacetime armament practice camp (APC). After an enjoyable and profitable month it returned to base to find that in its absence the High Speed Flight had been formed at Tangmere under Gp Capt Teddy Donaldson. This special unit, flying Meteors under the former No 1 Squadron Flight Commander, managed to raise the World Air Speed Record to over 616mph during the following month.

On 14 September No 1 was selected as the leading squadron in a flypast of 300 fighters over London and southeast England. On the following morning they lined the Spitfires tail to tail on the tarmac and bowed their heads at a drumhead service in memory of those who had not come back.

Two days after Battle of Britain Sunday it was learned that No 1 was to convert to the Meteor Mk III during the following month. The first of the new twin-jet fighters arrived on 7 October, and Sqn Ldr Dizzy Allen flew one for the first time that afternoon, coming back bubbling with enthusiasm at is performance. He found that the standard 'Meatbox' could attain 500mph at 30,000ft, which it reached at a rate of 4,000ft/min — well in excess of anything a piston-engined fighter could offer.

Later in October, Sqn Ldr Allen was posted to India to take over No 20 Squadron at Agra, arriving to discover that it had been disbanded in June! His place as CO of No 1 was taken by Sqn Ldr Colin Macfie DFC.

After a brief APC at Acklington in March 1947 the Squadron was deployed to Lubeck in Germany for a major exercise which lasted throughout May, then during July Sqn Ldr T. R. Burne DSO, DFC took over from

Below:
WO Watson brings his Meteor 3 alongside the Squadron's Oxford for a photo in December 1946.
Squadron Archives

Bottom:
The shape of things to come. Flt Lt Brand, the Engineering Officer, explains the finer points of the Meteor's Derwent engine. *Squadron Archives*

Sqn Ldr Tommy Byrne with some of his pilots at Tangmere in 1948, shortly after the squadron was re-equipped with the Meteor 4. *Squadron Archives*

Maj Robin Olds, the first USAF exchange officer to command a fighter squadron in the RAF, with his actress wife Ella Raines. *Squadron Archives*

Sqn Ldr Macfie. Soon after his arrival a very terrible thing happened: the Meteors were taken away and replaced by Harvards and, worse still, Oxfords! No 1 was to become an Instrument Flying Training Unit. This lamentable period lasted from 11 August 1947 until 1 June 1948, when No 1 once more became a front-line fighter squadron. During their 11 months of purgatory, Tommy Burne and his five instructors passed out over 150 rated pilots under the RAF's new Instrument Rating Scheme.

The Squadron's first Meteor F4s arrived at Tangmere at the end of May 1948. The summer was spent in converting new pilots, and No 1 became operational once more at the beginning of August.

The Squadron's first USAF Exchange Officer arrived at Tangmere in October 1948 to command 'A' Flight. This was Maj Robin Olds, who had been one of the top American aces of the war and who soon emerged as one of the dominant personalities of the Squadron. He achieved the highest scores at the APC at

Maj Donovan F. Smith USAF commanded No 1 from January to August 1950. During his time, RAF squadrons were allowed to re-adopt the colourful markings of the prewar era. However, the Meteor 4 in the background proved to be rather too flamboyant for the authorities to accept.
Squadron Archives

Acklington in January, and his instruction was instrumental in bringing the Squadron's average air-to-ground gunnery score up to 41% during the following month. It was no surprise that when Sqn Ldr Burne was posted in March, the Air Ministry approved the appointment of Robin Olds to command No 1 for the rest of his tour. This was the first time than an exchange officer had commanded an RAF squadron in peacetime.

The Tangmere Wing was detached to Manston on 23 June 1949, to take part in Exercise 'Foil', the major air defence exercise of the year. In simulated attacks on invading B-29s, Lancasters, Lincolns and Mosquitos the Squadron claimed 60 aircraft destroyed and 39 damaged. Claims were confirmed by examination of gun-camera film, and No 1 was the top scoring fighter unit of the Exercise.

In October Sqn Ldr Tommy Burne was posted back to No 1 as a 'caretaker CO' until the arrival of another American exchange officer, Maj Donavon F. Smith, who took over in December 1949.

The Squadron entered a four-man aerobatic team in the Fighter Command Championships in 1950, but they were withdrawn in April after the leader, Flt Lt P. W. Speller, was killed in a mid-air collision over Tangmere. The tail of his aircraft was cut off by one of the other Meteors, and the wreckage fell in front of the spectators near the control tower.

Another tragedy occurred a month later, when Plt Off R. W. M. Dixon was killed in a landing accident while practicing an asymmetric overshoot. This was a fairly common occurrence in Meteor units at the time, and many RAF pilots lost their lives in this way. Dixon had only been with the Squadron for two weeks.

When Maj Smith was posted home in August 1950 the command of the Squadron passed to Sqn Ldr J. L. W. Ellacombe. On the day after the new CO took over, the Squadron received the first of its new Meteor F8s, equipped with uprated Derwent engines of 3,800lb static thrust. In addition, a redesigned tail unit gave the new type better handling characteristics at high Mach numbers. Soon afterwards, No 1 was deployed to Biggin Hill for the first two

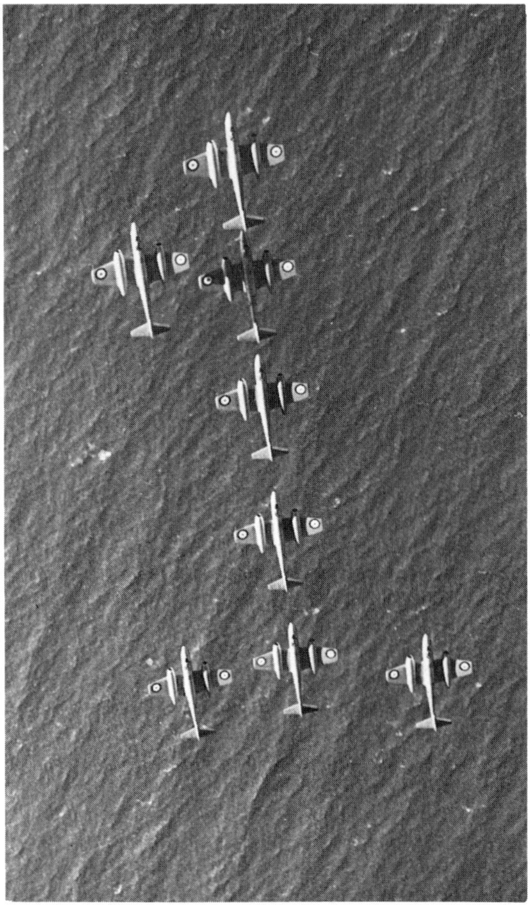

The 'Flying One' formation of Meteor 8s, taken by John Yoxall of *Flight* to repeat his earlier photograph of the Squadron's Furies. Yoxall was again flying in Hawker Hart G-ABMR. *Flight*

weeks of October to take part in Exercise 'Emperor', during which AVM Tom Pike, AOC No 11 Group, spent two days with the Squadron and scrambled on several sorties in one of the Meteors.

The Squadron's establishment had by now been increased to 20 pilots and 12 aircraft. The monthly flying task was set at 360 hours and was often comfortably exceeded during Sqn Ldr Ellacombe's tour.

Until the introduction of the Canberra into Bomber Command, the Meteor squadrons had found the art of high-level interception to be relatively easy. But the new jet bomber proved to be a much tougher proposition than its piston-powered predecessors. The efforts of the Meteor pilots to incercept Canberras on exercises were usually fruitless, and Flt Lt Evans was heard to remark that his throttles had been so far forward that he'd had to unstrap to reach them.

A number of young pilots were now beginning to arrive on the Squadron direct from training to fly the Meteor on their first tour. One of these was Plt Off M. R. T. Chandler, from Kenya, who had joined No 1 at the end of 1951. Mike's hobby was racing light aircraft, and he owned a Magister, which he entered in the Norton-Griffiths trophy race in July. He was approaching the finish in second place when he flew through a row of high-tension cables; there was a blinding flash as he hit them, but the aircraft stayed in the air long enough for him to cross the finishing line without losing a place. The Magister was too badly damaged to compete in the forthcoming Kings Cup race, but Chandler was lent a Miles Hawk in which he took high places in both the Kings Cup and the *Daily Express* South Coast races.

Sqn Ldr Ellacombe handed over No 1 Squadron to Sqn Ldr R. B. Morison DFC in October 1952. During the same month it was announced by the Air Ministry that the Squadron's Standard would be presented during the following spring. The award of the Standard had been granted by King George VI in September 1943, when No 1 had been flying Typhoons from Lympne, and as the Squadron was the senior of some 30 units waiting for their standards it was only proper that it should be the first recipient.

The presentation parade took place at Tangmere on 24 April 1953. A sunny morning saw 12 highly-polished Meteors drawn up on that same Tangmere tarmac where so much of the Squadron's history had unfolded, Sqn Ldr Morison led No 1 on to the parade ground and took post in front of the Standard, which lay in its case on piled drums. The Standard party, consisting of Fg Off Chandler with FS McNab, FS Glover and FS Scott, marched forward and stood while the banner was consecrated by the Chaplain-in-Chief. Chandler then knelt while the Standard was presented by AVM Sir Charles Longcroft KCB, CMC, DSO, AFC, who, nearly 40 years previously, had re-formed No 1 Squadron as an aeroplane unit at Farnborough. After a short speech by Sir Charles, the Standard was marched past in slow and quick time.

In addition to Sir Charles Longcroft, many distinguished former members had come to Tangmere to watch their old squadron receive this great honour. The senior serving officer

Above:
Meteor 8s at Tangmere in 1952, when the Squadron was commanded by Sqn Ldr J. L. W. Ellacombe DFC.
Squadron Archives

Left:
On 24 April 1953 AVM Sir Charles Longcroft presents the first squadron standard in the RAF to Fg Off M. R. T. Chandler.
Squadron Archives

was AM Sir Dermot Boyle, now the Commander-in-Chief of Fighter Command. Standing beside him were ACM Sir Philip Joubert, AVM Theodore McEvoy, Air Cdre Roy Chappel, Gp Capts Eustace Grenfell and Johnny Walker, Wg Cdrs Prosser Hanks, Billy Drake and Dizzy Allen, to name just a few. All those still serving had made a point of arriving by air — four of them flying their own Meteors. It was a day of which every member of the Squadron could justly be proud.

No 1 Squadron was detached to Biggin Hill on 27 April to take part in the Coronation and Royal Review flypasts. The Coronation Flypast over Westminster took place in very bad weather on 2 June, but was a complete success in spite of the conditions. The Queen's Review of the RAF took place at Odiham in much better weather. This time, No 1 Squadron was leading the first wing of fighters in the flypast and after landing back at Biggin Hill, the pilots were gratified to hear their AOC, AVM The Earl of Bandon, announce that the timing of the flypast had been 'near perfect'.

Number 1 returned to Tangmere in mid-July to resume its normal training tasks. While the Squadron had been away at Biggin, No 602 (City of Glasgow) Auxiliary Squadron had been spending its summer camp at Tangmere and its officers were dined out of the Mess a few days after No 1's return. It was a noisy, successful evening and No 1's pilots were, as usual, the last to leave the Mess. But they found to their dismay on the following morning that the Scots had pinched an important trophy — a top hat that had once belonged to the doorman at Claridges. Honour was duly satisfied by stealing one of No 602's Vampire FB5s, repainting it in No 1's colours and using it as the centre-piece for a Squadron photograph.

Sqn Ldr Morison was promoted at the end of July and posted to Stradishall as OC Flying. His place was taken by Sqn Ldr D. I. Smith, who spent a short and successful spell in command before falling ill in December, and being replaced by Sqn Ldr F. W. Lister DSO, DFC. Freddie Lister soon became a very

No 1 Squadron's first Standard — showing the original eight battle honours. *Squadron Archives*

While on detachment at Biggin Hill for the royal flypast, Fg Off Paddy Hine bought this taxi. It conveyed No 1's pilots on many memorable expeditions to the West End and around Tangmere; its owner had great difficulty in finding an insurance company which would cover 14 fighter pilots to drive it. *Squadron Archives*

The Squadron aerobatic team in November 1953, led by Plt Off Nick Nicholls, with Fg Off Paddy Hine flying No 2, and Fg Off Mike Chandler, from Kenya, at No 3. *Squadron Archives*

popular CO. He was quite a bit older than most of his officers but was a very experienced pilot and a superb leader in the air. He was also a great party man who considerably enlivened Tangmere's social life.

No 1 led another Royal Flypast in May 1954. For this occasion the Squadron formed a Wing with No 41, its arch-rivals at the time, to welcome the Queen back from her first overseas tour as Sovereign.

Four of the Squadron's more experienced pilots were posted out during August to help re-form No 34 Squadron, which was to join Nos 1 and 29 in the Tangmere Wing. This addition of a further day fighter unit at Tangmere could do nothing but good, for 29, splendid chaps though they were, flew Meteor night fighters and carried navigators. As such, they were quite beyond the pale as far as No 1 was concerned.

Over the years the Squadron had frequently

been engaged in low-level interception exercises along the south coast. These provided excellent flying but they had tended to upset the local populace, whose peace was frequently disturbed by dogfighting Meteors flashing past their chimney pots. No 1 was not the only guilty party, but those citizens of Sussex who took exception to the sound of freedom naturally tended to blame the best known squadron in the area. Soon, angry letters began to appear in the press. One, from a farmer who had been innocently 'contemplating his bullocks in a field', complained that a Meteor had set his livestock careering over the landscape. Other letters followed in a similar vein, and the local papers took up these grievances under dramatic headlines such as 'Banshees of the Air', 'Did Speed Bang Move the Wall?', etc. The matter was climaxed by a worthy councillor of Shoreham, who claimed at a Council meeting:

'We all know what it is. They have a couple of pink gins, take up a "kite" and proceed to have a "do". We are the people who suffer. Let us protest and stop the fun'.

Such allegations brought pressure on the Air Ministry to impose more low-flying regulations, and the number of complaints soon fell at the expense of combat proficiency. Several correspondents from the local press were flown in No 1's two-seat Meteor T Mk 7 and afterwards wrote heroic articles about their experiences, stressing that the pilots avoided overflying villages and towns wherever possible.

By the autumn of 1954, it was obvious that the Meteor was completely out-dated as a front-line fighter because it was incapable of intercepting modern bombers at high altitude. Therefore the news that No 1 was soon to be re-equipped with the Hawker Hunter was received with great joy of Tangmere.

The first sign that the Squadron was moving out of the Bronze Age was the issue of protective helmets — known ever since as 'Bonedomes'. Their introduction was not universally popular, for the pilots found that the new helmets cut down their rearward vision in combat. But they were made compulsory, so the boys sat down to paint them up in Squadron colours. This was promptly forbidden on the grounds that the non-inflammabale surface of the helmets would be ruined by amateur efforts at decoration. The opinion of the Squadron was that if your head was on fire it was definitely time to eject.

The Squadron's first Hunter, a Sapphire-engined Mk 2, arrived at Tangmere early in

The new squadron marking on a Hunter 5 at the end of 1955. The marking has maintained this shape ever since, but it has been steadily shrinking over the years. *Squadron Archives*

February 1955. It was immediately set upon by hoards of men brandishing spanners and screwdrivers, and it did not fly again for many weeks. Then, in March, both of the Flight Commanders, Flt Lts Harry Irving and Freddie Pickard, went to West Raynham to take the Day Fighter Leaders Course. There they were converted to the Hunter Mk 1 in order to spread the swept-wing gospel when they returned to Tangmere.

Conversion to the Hunter was delayed until June, when it was decided that No 1 should be the first unit to receive the new Mk 5. This model, like the Mk 2, was Sapphire-powered, but was equipped with an uprated engine and provision for underwing stores. Like the earlier models the Mk 5 was supersonic in a dive, and Wg Cdr Peter Simpson, the OC Flying at Tangmere, was the first to join the 'Mach Busters Club' when he dropped a most satisfactory sonic boom over Tangmere at the end of June. In this he was followed by Sqn Ldr Lister and the other pilots as they converted to type.

The conversion to the new aircraft was completed by the end of July. Although the Hunters were initially plagued with teething troubles, which did not endear them to the groundcrew, the pilots' opinion was unanimous: the Hunter was the finest and most exciting aircraft they had known — a view shared by generations of Hunter pilots the world over.

The Squadron got its first chance to show off the Hunters in September, when eight pilots took part in Fighter Command's 64-ship flypast at the SBAC display at Farnborough. Following this, Tangmere was 'at home' to the public a week later for the Battle of Britain display, for which Sqn Ldr Lister led a formation team of seven aircraft, while Harry Irving gave a polished solo aerobatic show. No 1's long association with Sydney Camm's flying machines had been pleasingly and publicly renewed.

During the winter a number of more serious technical faults cropped up to hamper the Squadron's operational training. Plt Off Turner's aircraft suffered a main undercarriage leg failure when landing on 28 October and narrowly missed the GCA radar caravan. Turner stepped out of the wreck unharmed, but investigation revealed that six other Hunters had cracks in their undercarriages and had to be withdrawn for repair.

Fg Off Alan Ginn had a more catastrophic accident on 5 December when his tail structure collapsed while diving at near supersonic speed. He ejected immediately but his legs were badly injured by flailing in the slipstream. The accident was due to a malfunction of the tailplane control system and caused yet more aircraft to be grounded for modification.

These problems caused the CO to go cap in hand to No 34 to borrow a couple of its Hunters. Fg Off Derek Bryant then caused a minor rift when he baled out of a No 34 Squadron aircraft after an engine failure.

The technical situation improved considerably during the spring of 1956, and May saw the completion of an intensive gun firing trial. The Hunter had been plagued with firing problems, but a series of modifications had now taken effect just in time for the Tangmere Wing to stand by for the Suez Crisis. Egypt's President Nasser was threatening to take over the Canal and a major international row was brewing.

The Tangmere Wing was now tasked to give fighter cover from Cyprus in the event of any British military intervention. As Hunters had never been to Cyprus before, a proving flight of four aircraft was sent out during May, led by Wg Cdr Simpson. They flew to Cyprus by way of Istres, Malta and El Adem, causing considerable interest wherever they landed.

The training programme was intensified at Tangmere, and the Squadron flew 240 hours in June: this was almost double the previous best with the Hunter and indicated the end of its teething troubles. Most of the training was slanted towards low level combat and interception, as befitted the current situation. In June Lt Mike Brown, a naval exchange pilot, pulled off a most commendable forced landing at Chivenor when his engine failed over low cloud. He was brought down by radar to within a mile or so of the airfield, in a position from which he decided to land on a disused runway, which was blocked at the time by a couple of taxying Hunters. He landed short of them at high speed, hauled his aircraft over the top and touched down again on the far side. Unfortunately he ran out of brakes and wrote off his wheels on the sea wall, before coming to rest in the Taw estuary, but suffered only a broken ankle.

Sqn Ldr Freddie Lister was posted to Germany at the end of June after a long and happy tour in command of No 1 Squadron. On the morning of his departure he was presented with a silver Hunter contributed to by every member of his unit — an indication of the tremendous respect and esteem in which he had been held by all his men. His designated successor failed the DFLS course, so Flt Lt

Above:
Hunter 5s at Tangmere in July 1956, shortly before the Suez crisis. *Squadron Archives*

Bottom:
The Tangmere Wing at Akrotiri in Cyprus at the start of Operation 'Quickfire', during which No 1 Squadron provided top cover over the landings in the Suez Canal zone. The aircraft in the immediate background are from No 34 Squadron — No 1's are farther up the line. *Squadron Archives*

Harry Irving was installed as acting CO until a new one could be found.

Nos 1 and 34 Squadrons were deployed to Akrotiri on 7 August, after the Wing had been standing by for the move for several weeks. All 25 aircraft reached Cyprus exactly on time and both Squadrons were combat-ready within 36 hours of arrival. Sqn Ldr R. S. Kingsford arrived in Cyprus to take over from Harry Irving in the middle of August, finding his Squadron to be in good spirits and ready for anything, although little was happening on the operational front. It was now obvious that the Wing would be on the island for much longer than the planned 10 days, and additional supplies and spares had to be sent out from England.

A unit of French RF-84s arrived at Akrotiri at the end of August. Accommodation was limited, and because the 'lead-sleds' needed all the runway they could get, the Tangmere Wing was moved north to the smaller airfield at Nicosia on 1 September. Both airfields were now completely crowded: in addition to the Venoms and RF-84s at Akrotiri, Nicosia began to fill up with several UK-based Canberra and transport squadrons. The tiny messes at

Nicosia could hardly cope, and all the visitors were accommodated under canvas — that is, until some enterprising pilot suggested that the Tangmere Wing's aircrew be moved into the squash courts. This was all done very properly. Each squadron had one court; the ordinary pilots slept on camp beds down on the floor, while the four Flight Commanders shared the balcony between them. On the first night in the courts, two resident Army officers were highly annoyed to be deprived of a game, and threatened all sorts of legal action before beating a hasty retreat in the face of 30 derisive pilots.

For the first two months at Nicosia the Squadron flew a variety of air defence exercises against the Venoms from Akrotiri, and also mounted a Battle Flight to investigate unidentified aircraft. On one occasion a DC-6 in French markings was intercepted over the island and 'invited' to land. Its pilot had no intention of so doing, but was kept circling for several hours by relays of Hunters until he ran out of fuel and was obliged to land at Nicosia.

Two flag-showing flights were made to Jordan during October. There was a certain amount of civil unrest, and the Hunters were sent as a gesture of British support for young King Hussein. On the first of these visits Sqn Ldr Kingsford and Flt Lt Jimmy Mansell provided an air display, while a week later Fg Offs Derek Bryant and Colin Shaw were presented to the Jordanian king.

Finally, it all happened in November, when the nationalisation of the Suez Canal was at last challenged by the British and French Governments, which ordered an airborne assault on Gamil airfield and the destruction by Venoms and Canberras of military installations along the canal.

The task of the Tangmere Wing was to provide fighter cover for the paratroop transport aircraft and to prevent Egyptian Air Force intervention. The drop on Gamil airfield took place on 5 November, and while the Valletas and Hastings were in the target area they were covered by relays of four Hunters from both squadrons of the Wing. The pilots had a ringside seat for all the action, but the anticipated reaction of the Egyptian MiGs never materialised.

During the following two days the Wing conducted several fighter sweeps at 30,000ft over the Nile Delta, once penetrating as far south as Cairo. On one occasion they did see a single MiG-15, but the pilot wisely flew in the opposite direction at high speed.

The Hunter had fuel for only 10 minutes in the target area, so the Wing was withdrawn from the offensive action after the first few days and was tasked with the air defence of the Cyprus bases. This resulted in some rude ribbing from the Venom pilots from Akrotiri, who had seen considerably more action over the Canal. Nevertheless, the aircraft parked at Nicosia had another threat to contend with. At the end of November, No 1 lost a Hunter when a bomb planted by EOKA terrorists exploded in the wheelbay of WP180. This resulted in a large increase in guard duty for the groundcrew, and the pilots' pre-flight inspections of their aircraft suddenly became very painstaking!

Pilots outside the Squadron operations tent at Nicosia, during Operation 'Quickfire'.
Squadron Archives

The Squadron left Nicosia on 21 December 1956 to return to Tangmere. The Hunters were delayed at Malta by fog in England, but they arrived on Christmas Eve and were joined at Tangmere by the groundcrew later in the day. It had all been good experience, but no one was sorry to be home.

A most enjoyable squadron exchange was completed during February when six of No 1's Hunters, led by the CO, flew to Cambrai in France. While they were there the 1st Squadron of the 12th Wing of the Armée de l'Air sent six of its Mystere IVAs to visit Tangmere. Both the home and away fixtures were a great success, and although bad weather prevented much inter-squadron combat flying taking place, the CO and his boys were taken by their hosts for an excellent dinner at the Crazy Horse Saloon in Paris. Similar festivities were meanwhile taking place at Tangmere: the 'Spotted Cow' at Chichester suddenly became 'La Vache Tachée', and the Entente Cordiale was measurably strengthened on both sides of the Channel.

Life at Tangmere continued much as before throughout the summer of 1957, with flying devoted mainly to high-level battle formation, interceptions and air-to-air combat. The summer exercises 'Vigilant' and 'Fabulous' passed without incident, and at the end of August the Squadron was deployed to Biggin Hill for a fortnight to take part in the Farnborough flypast. Shortly before Christmas, No 1 sent eight Hunters to Holland for another squadron exchange visit, this time with No 327 Squadron, RNethAF, at Soesterberg near Amsterdam.

The links between No 1 and the people of Brighton were renewed most successfully at a 'friendship weekend' in January 1958. As well as an ice show on 25 January and a church parade on the following morning, the pilots attended a ball in their honour at the Metropole on the Monday night. It was unfortunate that the Air Ministry should have chosen that morning to announce the forthcoming closure of Tangmere as a fighter station! The news rather took the edge off what was otherwise an excellent party, but Sqn Ldr Kingsford went to some pains to point out that No 1, wherever it was posted, would never forget its links with the townspeople of Brighton.

The impending demise of England's finest fighter station was one of the many unfortunate results of the Defence White Paper of 1957 — known throughout the RAF as the 'Sandys Axe'. In his White Paper, Mr Duncan Sandys told the world that the last manned fighter would be the Lightning, and that missiles would thereafter replace pilots. As a result of this policy statement, over 30 fighter squadrons both at home and in Germany were disbanded.

It seemed that those squadrons flying the early marks of Hunter would be among the first to go, and this, coupled with the impending closure of Tangmere as a fighter station, was ominous news for No 1. The situation looked even blacker at the end of January, when No 34 Squadron was disbanded and re-formed with Blackburn Beverleys!

Eventually the news arrived that No 1 itself was to be disbanded, though fortunately it was to be re-formed simultaneously at Stradishall by renumbering No 263 Squadron, which had recently re-equipped with the Avon-powered Hunter F6.

The pilots and groundcrew of No 1 Squadron were to be dispersed when disbandment occurred, but to maintain continuity two pilots, Fg Offs Pete Highton and Bob Turner, were to be transferred to the 'new' No 1 at Stradishall. The date of the handover was set at 23 June.

The members of the 'old' Squadron at Tangmere made the best of that last spring and summer at No 1's ancestral airfield, which, with its beaches and pubs, its fine flying and sea air, would soon be just another memory for the greatest of its old squadrons. The final month at Tangmere, with all its parties and festivities, was marred by a tragic accident when Flt Lt Michael Paxton and Fg Off John Turner failed to return from a sortie at the end of May. It was thought that they had been involved in a mid-air collision, and a memorial service, the last of so many at Tangmere, was held in the station church on 29 May.

The Squadron threw a number of parties before it left Sussex. A dinner dance in Brighton on 13 July was attended by many familiar faces who had served with No 1 during the past three years, the all-ranks dance at Chichester was a great success, and so was Sqn Ldr Kingsford's barbecue held on the 19th. During the barbecue a series of shattering explosions was heard close overhead; someone had tied a number of thunderflashes to a met balloon and sent it drifting over the party.

The Squadron paraded the Standard at Tangmere on 23 June. The salute was taken by ACM Sir Thomas Pike as 12 aircraft flew past in immaculate formation, and Fg Off Pete Highton then handed the Standard of No 1 Squadron to Fg Off Anthony Mumford of No 263.

The handover symbolised the end of an era. Tangmere, with all its nostalgia and history, would never more echo to that evocative blue note of Meteors and Hunters running home to break and land. Never again would the bars of the Old Ship, the Spotted Cow or the Red Lion at Pagham echo to the merry laughs and bawdy stories of fighter pilots. It was a sad day for Sussex and for No 1.

CHAPTER SIXTEEN

Ground Attack

No 263 Squadron assumed the mantle of No 1's fine past at Stradishall on 1 July 1958, the parading of the Squadron's Standard finally cementing the change of identity on 22 July. The new No 1 was a worthy inheritor. Commanded by Sqn Ldr L. de Garis AFC, it had been one of the crack units of Fighter Command for many years.

Remaining at Stradishall, the Squadron was combined with No 65 at Duxford to form No 165 Wing. Both units had recently been equipped with the Hunter F6, a new model powered by an uprated Rolls-Royce Avon of 10,000lb thrust, which gave it the nimblest performance of any Hunter. Like its predecessors, the Mk6 was armed with four 30mm Aden cannon, but its wings had been strengthened for extra fuel tanks and weapons on four hard points. Provision had also been made for fitting up to 24 3in rockets on underwing rails.

No 1 led a formation of 45 Hunters over Farnborough at the annual SBAC display in September. As the six squadrons involved were all based within striking distance of Cambridge, Sqn Ldr de Garis arranged an inter-unit punt race on the Cam to mark the occasion. The Squadron's team was finishing strongly when they were caught by a broadside of missiles from their rivals, but Capt Ted Hassig, No 1's Dutch exchange officer, refused to admit defeat. His punt was sinking fast, but he fought a Canute-like battle with the waves, continuing to paddle furiously as the waters closed over his head. The team bought the beer with good grace.

The Squadron on the tarmac at Nicosia in October 1958. During this deployment No 1 flew several sorties in support of British forces operating against EOKA terrorists. *E. J. Soons*

At the end of September, No 1 took 14 Hunters to Cyprus to convert to the ground attack role. During the month at Nicosia, the pilots flew an intensive programme of low-level simulated strikes on ground targets and live attacks on Larnaca range, with some spice added to the detachment by EOKA terrorists who were ambushing British convoys in the southwest of the island. The Hunters were tasked to fly top cover to a number of these convoys and on one occasion came to the rescue of a Grenadier Guards patrol which was pinned down by sniper fire from a village belfry. Flt Lt Tom Gribble and Fg Off David Brook made a couple of dummy passes on the belfry and the gunmen ran away. Also, a couple of British reporters who were driving to the scene collided with a Turkish roadblock when the Hunters flashed past their wind-screen at very low level.

After returning to Stradishall, Sqn Ldr de Garis handed over to Sqn Ldr J. J. Phipps shortly before Christmas 1958.

The spring of 1959 was enlivened by an incident which befell Fg Off Brook on 17 February. The fog was rolling in at Stradishall and he hit a tree while approaching to land in very poor visibility. The nose and wings of the Hunter were badly damaged, but he found that he could still keep it in the air. However, by the time he had completed a low-speed handling check the fog had closed in completely, and he was sent off to nearby Honington with only 200lb of fuel, where he pulled off an excellent downwind forced landing with the aid of the base's CRDF radio homer. After his forcible hint the offending tree was completely removed by properly qualified woodsmen.

Four of the Squadron's aircraft led the Queen's birthday flypast over Buckingham Palace on 13 June; they arrived precisely one second early. Two weeks later the annual inspection of Stradishall by the AOC No 11 Group took place. After watching a demonstration scramble and operational turn-round, the AOC gave permission for the march-past, and, to his amazement, a large balloon in the Squadron's colours of red and white appeared round the corner of a hangar. It was towed by 17 'engines' — eight officers, eight airmen and the Flight Sergeant — with the Adjutant, Anthony Mumford, standing in the basket. Mumford gave the order 'eyes right' as the cavalcade passed the AOC, but his smart salute turned into an undignified lurch as the balloon decided to dip rapidly to 20ft: the 17 'engines' were too busy to hear the AOC shouting at them to cast the Adjutant adrift. Fg Off Mumford was later granted a green rating 'for valour' on balloons by the AOC. (The balloon had been brought to Stradishall by Mr Gerry Long who had served with the Squadron on its formation in 1912.)

No 54 Squadron moved to Stradishall during July and was formed with No 1 into a new Wing — No 154 — under Wg Cdr A. D. Boyle. This union was the start of another period of fierce and friendly rivalry for both squadrons.

The *Daily Mail* Bleriot Anniversary race from London to Paris also took place during July. Some weeks previously, Sqn Ldr Phipps had submitted proposals for a Squadron team to enter the race, using a two-seat Hunter T7 for the transit, with helicopter and RAF police motor-cycle lifts at either end. He was annoyed when his plan was adopted for use by a team from No 65 Squadron, and it was only some consolation that one of the T7s was flown by a No 1 Squadron pilot, Flt Lt Roly Jackson. He flew nine times between Biggin Hill and Villacoublay as taxi driver to various members of the team, with an average flight time of 18¼ minutes. The runs were made at full throttle at 2,000ft, which gave the Hunter a speed of about 0.92 Mach, and this left very little margin for error in poor visibility while using the minimum possible fuel state. The Squadron's T7 (XL601) was badly twisted when another pilot dropped the undercarriage at over 300kt during a straight-in approach. She was still flying sideways at Wittering in 1971!

During the autumn both squadrons of the Stradishall Wing began to receive their first ground-attack Hunters. These were interim aircraft, Mk 6s which had been partly modified to the standard of the projected F(GA)Mk 9. Their wings had been further strengthened to carry 230gal drop tanks on the inboard pylons, and a braking parachute had been fitted.

By the end of the year the Squadron's pilots had been checked out to fire the 3in rocket which had been used to great effect by Typhoons and Tempests towards the end of

Above right:
A Hunter 9 firing 3in rocket projectiles at Cowden range on the Yorkshire coast. These rockets had changed little since World War 2 — they were normally fired by the Hunter at 400kt in a 30° dive. *Squadron Archives*

Right:
Four Hunter 9s in box formation over Kyrenia Harbour in Cyprus during the summer detachment of 1960. *Squadron Archives*

Flt Lt John Quarterman making a moderately low pass at El Adem range, near Tobruk in Libya, in May 1962. Between 1962 and 1969 No 1 visited RAF El Adem each year for its Armament Practice Camp and for desert exercises with the Army.
Wg Cdr J. Quarterman

the war. After a demonstration in the T7 from the Pilot Attack Instructor (PAI), each pilot would be let loose on the Holbeach range with four of the '3in drainpipes' fitted with 60lb concrete practice heads. By the end of December they had all fired on several sorties and had recorded an average miss-distance of 17yd — not bad for a first attempt.

The Squadron visited Cyprus again in the summer of 1960, the 10 aircraft fitted with 230gal drop tanks on the inboard pylons and 100gal tanks outboard. It was No 1's first long range ferry flight in this configuration, and all aircraft arrived on time at Nicosia on 1 June. During the seven weeks in Cyprus they concentrated on gunnery. Flt Lt Harry Davidson distinguished himself on 16 June, by scoring 100% hits on the banner towed behind a Meteor. A total of 58 holes were counted in the target, even though Flt Lt Davidson had only fired 50 rounds — the other holes were due to the flag rippling in the slipstream. He was selected as Captain of the Fighter Command Gunnery Team later in the year.

On 24 June four Hunters left Cyprus for a tour of the Middle East, flown by Sqn Ldr Phipps with Flt Lts Chuck Coulcher, Harry Davidson and Tony Chaplin. The latter had brake failure in Turkey and collided with the runway caravan when taxying to the dispersal, and one serviceable aircraft had to be left behind to pair up with the damaged Hunter, which had to be flown back to Nicosia with its wheels down. Sqn Ldr Phipps and Chuck Coulcher carried on and visited Sharjah, Khormaksar, Nairobi and Bahrain, before returning to Nicosia on 1 July.

No 1 returned to Stradishall at the end of July, having raised its strafing and rocketing averages to 50% and 16yd respectively. Flt Lt Mark Tomkins scored a direct hit with each of his four rockets on one sortie, while Fg Off Mac Cameron got 100% on an air-to-ground shoot.

Sqn Ldr John Phipps went to Staff College at the end of the year after an outstanding tour as OC No 1 Squadron. When he left he gave the Squadron a quart beer mug inscribed with the names of each of No 1's Commanding Officers, which has been put to good use since by his successors. He was replaced on 19 December, by Sqn Ldr P. V. Pledger.

The first two weeks of March 1961 were spent practising for the Buchanan Trophy, a silver tankard presented by Maj Buchanan, the Wing's Ground Liaison Officer, for the highest one-pass rocketing and air-to-ground scores between Nos 1 and 54 Squadrons. Sadly, Fg Off Barry Dimmock, a very popular young pilot, was killed when he crashed into the sea after a practice rocket attack at Cowden Range on 2 March; the first fatal accident on No 1 Squadron since the move to Stradishall. The depressing effect on his fellow pilots was shown in the results of the 'Buck Shoot': the average gunnery score was 12.5%, and even Flt Lt Davidson could only manage 29%. It was a hollow victory for a sympathetic Fifty-Four.

The Squadron went back to Cyprus in April, to take part in the joint-service exercise 'Fabulist'. It had been decided to try for the London-Valleta speed record on the way out, Wg Cdr Bennett (the OC Flying at Stradishall) and Flt Lt Harry Davidson making the attempt. They waited for a suitable tailwind until 25 April, and were rewarded with an average speed of 633.3mph, which beat the existing record (held by a Scimitar) by over nine minutes.

The APC of June 1961 was spent on the island of Sylt off the Friesian coast of Germany. Regrettably this was the only time that No 1 ever went to Sylt, as it was always the most popular of the RAF's air firing ranges. Clothes were an optional extra on some of the well-decorated beaches of the island, while at the others they were strictly verboten. The Squadron's increasing specialisation in ground-attack meant that this was to be its last air-to-air APC. From here on the camps were to be held twice a year at that other sunshine paradise — El Adem in the Libyan desert.

On its return from Sylt, No 1 was placed on three days' readiness to go to Kuwait, which was being threatened by the overtures of Gen Kassim of Iraq, but the Squadron was released when the situation cooled off on 11 July. Their planning came in useful, however, when four aircraft flew to the Persian Gulf in September, by way of Luqa, El Adem, Khartoum, Khormaksar and Sharjah. They arrived at Bahrain on 23 September and flew a sector recce of the Kuwait area, before returning to Stradishall on the 27th by way of Teheran and Nicosia.

No 154 Wing moved across Cambridge from Stradishall to Waterbeach in November 1961, to become the 'shadow wing' of No 38 Group of Transport Command. The role of the Wing was to provide tactical close air support for those mobile units of the army carried into battle by No 38 Group's transport squadrons. As a secondary commitment the Hunters were required to provide top cover for the Hastings and Beverleys, and Nos 1 and 54 were thus the only fighter squadrons ever to have served in Transport Command. The other Hunter squadrons were quick to let them know that they considered this to be a dubious privilege, but Nos 1 and 54 had the last laugh as they watched the others being disbanded one by one.

Wg Cdr John Phipps returned from Staff College at the beginning of 1962, to take over as OC Flying at Waterbeach. Soon after his arrival the Wing was visited by the Secretary of State for Air, the Rt Hon Julian Amery, MP, who after watching an operational turn-round of the Squadron's aircraft was flown in the T7 by Flt Lt Duncan Allison as part of a four-aircraft live rocketing strike on the Stanford Battle Area. Amery was generally very impressed with the proceedings, but was disappointed when an electrical fault in the two-seater prevented him from becoming the first Minister to fire a '3in drainpipe'.

No 1 completed its first APC at El Adem in May 1962. A successful firing camp was climaxed with an 'operational' shoot, when each pilot fired a salvo of four rockets with 60lb high explosive heads at some disused trucks set up in the range area — a most satisfying exercise. While in El Adem, the Squadron celebrated its 50th anniversay. The Hunters were lined up for the traditional tail-to-tail photograph and the champagne flowed freely. The permanent staff of El Adem did the Squadron proud. Wg Cdr Jay, the Station Commander, presented the Squadron with a magnificent birthday cake covered with iced replicas of the Hunter, and an excellent series of parties then took place to round off a very happy detachment. El Adem wasn't so bad after all!

The other big trip of the year took place in October when 10 of No 1's Hunters left Waterbeach for the Persian Gulf on 18 October. They were plagued with unserviceabilities all the way round the southern route, climaxed by Flt Lt Brian Scotford's engine failure over the Radfan mountains north of Aden. He tried to glide back to Khormaksar but had to eject 35 miles short of the airfield; he landed uninjured and was picked up without difficulty. The remaining aircraft stayed at Sharjah for a week to fire a few rockets and cannon shells into the Jebajib range, Flt Lt Tom Lecky-Thompson doing so to great effect, notching up the first 100% air-to-ground score for some time.

Four of the Hunters left Sharjah on 26 October for a three-day goodwill visit to Pakistan, and were led to Karachi by Flt Lt Mo Akbar, No 1 Squadron's Pakistani exchange officer. The remaining aircraft were flown by the CO, Duncan Allison and Plt Off Mike Perkins; Jerry Prag, the Engineering Officer, followed along behind in an Argosy with a party of the groundcrew. They flew on to Bahrain on the 29th, after a hectic round of parties and an immaculate farewell flypast. Mo Akbar, who did a great deal to make the visit a success, was the Squadron's second 'tame' Pakistani, having taken over from Flt Lt Mukhtar Ali in June.

Sqn Ldr Peter Pledger handed over to Sqn Ldr F. L. Travers-Smith when No 1 returned from Bahrain, and, promoted to Wing Commander, took over the post of OC Flying at Waterbeach from Wg Cdr Phipps. The latter went on to become a very young Group Captain, but tragically died very suddenly of a heart attack in October 1965.

No 1 Squadron was one of those units tasked with the rapid reinforcement of the flanks of NATO, as a part of what later became known as the ACE Mobile Force. Working closer with the Army, No 1 deployed to a tented camp on Kemble airfield in March 1963, to take part in Exercise 'Lifeline', during which the Hunters made mock attacks on transport aircraft and paratroop concentrations, working closely with Army Forward Air Controllers (FACs), who talked them in on their targets.

In August 1963 Nos 1 and 54 moved to West Raynham, to ease the crowded airspace around Cambridge. West Raynham, set in the quiet countryside of western Norfolk, seemed glad to greet the Hunters back to its empty circuit — the locals, who depended heavily on the station for employment, had feared that it would close when the Central Fighter Establishment had been moved away to Binbrook.

By 1964 the Squadron had become very proficient in the art of ground attack. The average miss distance with 3in rockets had fallen to 15yd and the average cannon score for all types of academic and pull-up attacks never fell below 40% for over two years. The Aden guns were fired in a 10° dive at 15ft-square hessian screens. By now, the Squadron had been cleared to fly some navigational and attack profiles at 50ft — which added a certain spice to life at 450kt: the significance of the 50ft clearance became

The officers of No 1 Squadron at Waterbeach in January 1963. Sqn Ldr Frank Travers-Smith is flanked by his Flight Commanders, Flt Lts Duncan Allison and Tom Lecky-Thomson. *Air Cdre D. Allison*

apparent at the beginning of 1965, when the pilots began skip-bombing with small practice bombs. These were delivered from a low level pass to simulate the dropping of napalm, whose use in peacetime was politically unacceptable, even for training.

Sqn Ldr David Brook returned to No 1 Squadron to take command from Sqn Ldr Travers-Smith at the end of 1964. Very soon after his arrival, General Franco began his propaganda campaign against Gibraltar, and the Squadron was ordered to deploy there with all speed on the evening of Friday 19 February 1965. Many of the pilots and groundcrew had already left West Raynham for the week-end, so it was a considerable achievement to get eight Hunters down to the Rock within 48 hours, the spare pilots travelling down with the groundcrew in four of No 38 Group's Argosies. The Squadron remained in Gibraltar for two weeks, continuing its normal flying training under the close scrutiny of the Spanish border guards. Nothing much happened during their stay, but Franco had witnessed a very rapid demonstration of British determination to protect the rock and its people.

The Hunters returned to Raynham by way of the Bay of Biscay, refuelling at St Mawgan in Cornwall en route. This was the first time that an over-water ferry of such length had been tried and it proved that No 1 could return if needed without the assistance of any foreign staging posts.

Two weeks after the return from Gibraltar, the Squadron was faced with a dramatic change of climate when 10 aircraft flew to Bodø in northern Norway to take part in the joint-service exercise 'Cold Winter'. This one really lived up to its name, for the long night of the Arctic winter was barely over when the Hunters arrived. The Norwegian pilots at Bodø had just received their Lockheed F-104s, and were quite startled when Flt Lt Jerry Seavers, the commander of 'B' Flight, won a combat against one of their new aircraft. This led to some excellent rivalry during the exercise, and resulted in a fearful series of parties before No 1 returned to England. A trail of black footprints meandered over the walls of the Officers' Mess at Bodø after the farewell thrash, the result of some brave gymnastics by the CO and Fg Off John Hough. Flt Lt Tommy Thomson, commanding 'A' Flight, went one better: he left the imprint of his bottom on the ceiling as a legacy of the Squadron's stay. For this exploit he became known as the 'Black Asstronaut'.

Tommy Thomson led four aircraft on a brief Ranger flight to Aden during April. Improvements in Middle Eastern relations permitted them to stage through Cairo and Jeddah before enjoying the hospitality of Nos 8 and 43 Squadrons for a few days. They then rejoined the rest of the Squadron at El Adem for the regular APC in May.

A rather unusual unit had meanwhile been formed at West Raynham. This was the tripartite Kestrel Evaluation Squadron composed of British, American and German pilots and groundcrew, whose task was to evaluate Hawker Siddeley's new Kestrel vertical take-off fighter — the first of its kind in the world. Those serving on No 1 at the time may not have realised the significance that this aircraft would have on their Squadron's destiny, but they lost no time in getting to know the pilots, and marvelled at the sight of a six-ton fighter standing still in the air. Nonetheless, it wasn't very long before one of the Hunters, flown by Fg Off Bob Iles, came home with some good gun-camera film of a Kestrel, just to remind the newcomers who the airfield really belonged to.

One of the German pilots on the Tripartite Squadron was Col Gerhard Barkhorn, who claimed to have destroyed the amazing total of 301 Allied aircraft during the war. One day Col Barkhorn had to make a forced landing in a Kestrel, which resulted in considerable damage to the aircraft. As he climbed out of the wreckage he is reputed to have muttered: 'Three hundred and two'.

Whitbread, the brewers, presented the Squadron with a magnificent inn sign on 16 July. This sign had hung outside the 'True Briton', a pub frequented by No 1 while stationed at Lympne in 1943, and depicted a woad-covered ancient Briton on one side and a Battle of Britain pilot on the other. The presentation, by Mr Thomson of Whitbread, also commemorated the consumption of unfathomable quantities of the company's excellent product by generations of No 1 Squadron down the years.

The unveiling of the inn sign was coupled with a ceremonial opening of the Squadron museum. This had been started in 1963 by Flt Lt Chas Boyack and had been gradually taking shape ever since. It contained some priceless relics of the Squadron's history, including the fins of a Fokker and a Pfalz shot down in 1918, and the gun with which Arno Frankenberger shot down 'Pussy' Palmer's Hurricane in November 1939.

The ceremony was attended by two distinguished former CO's of No 1 Squadron who presented the musuem with valuable additions. Maj W. E. Young gave some fascinating papers and photographs which dated from his time as CO in 1918, including the notebook in which he had roughed out the plans for all the daylight bomber escorts at the end of that war. Wg Cdr R. C. Wilkinson, CO of the Squadron at Lympne and former patron of the 'True Briton', also came to the ceremony. He presented the Squadron with a magnificent RAF No 1 dress of the 1920s, complete with busby. These, and the many other historical items in the museum have since been looked after with loving care by a succession of curators, and now adorn the Squadron's headquarters at Wittering.

But the pride of the museum has always been its Spitfire, LA255, coded JX:U, a Mk 21 which flew with the Squadron in 1945. Mr Leslie Hunt, the aviation historian, contacted the CO in 1964 to say that it was rotting away at Cardington. A rescue operation was mounted and the Spit was recovered to West Raynham on a Queen Mary transporter, to be renovated and parked snugly alongside the Hunters in the hangar. It has since been brought to Wittering to continue a well-deserved and pampered retirement.

The succession of overseas exercises and weaponry camps continued unabated throughout 1966. The Hunters of No 1 Squadron flew far afield to practise their art — Libya, Aden, Norway and Germany saw them come and go — and, in addition to major detachments, the mobility of the unit was further practised by formations of two or four aircraft going to airfields throughout Europe and the Middle East.

During the APC at El Adem in May 1966, Sqn Ldr Brook was challenged by the CO of No 32 Squadron that the Hunter pilots could not find the wreck of a crashed Liberator which had recently been found in the desert. *Lady be Good* had come down in the Libyan Sand Sea during the war, some 320 miles south of El Adem, on the way back from a bombing raid. She lay in the middle of one of the most featureless places on earth, yet the CO and Jerry Seavers found her after 40 minutes of faultless navigation. No 32 Squadron's Canberras took several sorties to find the wreckage, in spite of all their electronic aids

and professional navigators, so were obliged to stand No 1 an evening of drinks for their impertinence.

Tommy Thomson left the Squadron in 1966 and handed over 'A' Flight to Flt Lt Andy Whitson. Jerry Seavers was promoted to Squadron Leader in the July list, but remained on the Squadron until the end of the year in command of 'B' Flight. Two of the younger pilots on the Squadron, Flt Lts Mike Stear and Wally Walton, were detached to Chivenor for the three month course to qualify them as Pilot Attack Instructors.

Sqn Ldr David Brook finished his tour as CO in November 1966 and, as had long been the custom on No 1, he was dined out with due ceremony at the Officers' Mess at West Raynham before the arrival of his successor. Sqn Ldr Spike Jones arrived to take command on 19 November.

A good year always finishes with a bang. This was so for Flt Lt Peter Curtin on New Year's Eve. He and Ken Parry were returning from a gunnery sortie on the Cowden Range when Pete's engine started to fail. He tried to nurse his aircraft into Leconfield but had to eject just short of the runway when the Avon finally seized. The Hunter dived into a row of houses in Leconfield village, and blew the back out of a recently constructed rectory. The vicar, who was sitting on his loo at the time, was very surprised by the ensuing explosion, and suffered a broken arm. Pete collected a mild dose of 'Martin-Baker back' during the ejection, but was fit enough to travel back to West Raynham, where he saw the New Year in with some enthusiasm.

On 18 March 1967 the *Torrey Canyon*, a super-tanker carrying over 60,000 tons of crude oil from the Middle East, smashed into the Seven Stones reef near Land's End. Every attempt to tow her off the rocks failed during the week that followed and by the morning of the 27th the Atlantic rollers had broken her

Hunter 9s over the Lofoten Islands in northern Norway. From 1965 onwards, No 1 Squadron's commitment to NATO's ACE Mobile Force has led to frequent deployments to Norway and Denmark, with both the Hunter and the Harrier.
Squadron Archives

Flt Lt Jerry Seavers debriefing Fg Off Jerry Gosnell during the annual Buchanan Shoot competition against No 54 Squadron in 1965. The scoreboard shows that they have just given No 1 an early lead. Flt Lt Chaz Boyack looks on in approval.
Squadron Archive

back and she lay across the reef in three separate parts, spewing great slicks of crude oil which threatened to pollute every beach in the southwest.

The Government eventually decided to use air power to burn as much oil as possible before it left the tanks. A formation of Naval Buccaneers dive-bombed the wreck with great accuracy on the 28th and set it on fire, but relays of Hunters from Chivenor failed to keep the blaze going with 100gal drop tanks full of kerosene. The only weapon that could now be used to ignite the oil was napalm: the aura surrounding the 'Vietnam horror bomb' at the time was such that the decision to use it was taken at Cabinet level.

No 1 Squadron made its first napalm strike on 29 March. The three Hunters, flown by Sqn Ldr Jones, Flt Lts Wally Walton and Alan Pollock, loaded up at West Raynham and flew to Land's End by way of the English Channel, so avoiding carrying the firebombs over any populated areas. They hit the ship at 12.20hrs, Spike Jones making the first attack on the oil slicks surrounding the tanker. They failed to ignite. Wally Walton went in next. He flew down the centre-line of the *Torrey Canyon* and laid his napalm tanks at the base of the superstructure, where they blew up in a holocaust of smoke and flame. Wally was heard to shout 'Whoopee!' over the R/T as he climbed away and looked back over his shoulder.

Al Pollock was starting his run on the tanker when Wally's napalm went off. He broke off his attack when the smoke obscured the target, but made another run across the deck and dropped his tanks when the smoke enveloped his Hunter. All three pilots landed at Chivenor to refuel before returning to West Raynham in a state of great excitement.

The observers on board HMS *Barossa*, patrolling near the reef to keep the sightseers away, reported that the fires had gone out without igniting very much crude oil. Further napalm strikes were therefore flown during the afternoon, preceded by two Hunters firing salvoes of 12 3in rockets to open up the tanks. The rockets found their mark, but insufficient oil was spilt to guarantee ignition. Further bombing took place the next day, but operations were cancelled in the evening when it

Above:
The *Torrey Canyon*, taken through Flt Lt Wally Walton's gunsight, as he lines up for a napalm run on the stricken supertanker.

Below:
Sqn Ldr Spike Jones, back from the *Torrey Canyon* raid, interviewing some awe-struck reporters.
Squadron Archives

Fg Offs Ken Becker, Paul Wilcox and Mick Hindley debriefing on the grass at Tangmere with Flt Lt Alan Pollock after the RAF's 50th Anniversary flypasts at Brighton and Chichester. It was on the return sortie from Tangmere that Al Pollock flew his Hunter under Tower Bridge. *Squadron Archives*

was found that fires were becoming increasingly difficult to start.

Operation 'Mop-Up', as it was known, may not have saved the beaches from pollution, but it certainly resulted in an all-time high in Squadron morale. Most of the pilots had been involved in the strikes and all came home with grins from ear to ear. *Torrey Canyon* had been the perfect target: it was big, it burned well and it didn't hit back!

Continuing Spanish dissatisfaction with Gibraltar's status, resulted in the return of No 1 to the Rock in September, and for the following three years the West Raynham Wing had to maintain two aircraft and four pilots at Gibraltar to provide evidence of British support. The Gibraltar detachments had to be dovetailed in with the many other training tasks of both Nos 1 and 54 Squadrons, but were popular with the pilots, especially in the summer months. The inhabitants of the Rock were extremely friendly, the daily routine was leisurely and the social life was often very exciting.

The inter-squadron shoot for the Buchanan Trophy took place on 24 October, each pilot flying to the Theddlethorpe range at Saltfleet in Lincolnshire in company with a Hunter of the rival outfit. The aircraft carried two loaded cannon, two 3in rockets with concrete heads and two 25lb practice bombs. No 1's pilots established an early lead over No 54, which was handsomely increased throughout the day to result in a substantial and clear-cut win for No 1 Squadron. When flying finished for the day, the pilots provided two barrels of beer for their tireless groundcrew and then paid a musical visit on their defeated rivals. This took the form of the Squadron's string ensemble — known as 'Big Spike's Banjo Band' — mounted on a Land-Rover and trailer.

No 1 deployed to El Adem on 1 November for its regular APC. By now the inventory had been increased to include the 68mm SNEB anti-tank rocket. This high-velocity weapon was fired in salvoes of 18 from a pod carried on the

Hunter's outboard wing pylons and it soon replaced the old '3in drainpipe'.

During the APC, a night ground attack trial took place which was viewed by the pilots with some alarm, but target illumination was provided by flares dropped from Canberras and the exercise proved to be less terrifying and more successful than had been imagined. A truly frightening incident was suffered by Flt Lt Paddy Hughes who had a medical blackout during a 30° rocket dive and only regained consciousness at low level, several miles from the range and heading for Egypt. He landed successfully but later retired on medical grounds.

When off-duty the pilots would gather at the Tobruk beach club. Someone remarked to Wally Walton about the size of a large tanker that was berthing at the oil terminal across the bay. Wally replied truthfully that he'd set fire to a much bigger one than that.

The 50th anniversary of the RAF was celebrated on 1 April 1968. No 1 being almost 57 years old was initially inclined to view the whole proceedings with an air of amused tolerance. As the date of the anniversary approached, however, it appeared that the whole occasion was being played down, and many people throughout the RAF felt hurt and upset that the Service was not to be permitted to mount an anniversary flypast over London. No 1, with all its tradition, felt that some gesture should be made.

Soon after the anniversary parade at West Raynham was over, a Hunter roared over 54's dispersal at very low level, scattering birthday leaflets and lavatory paper from its flaps and airbrakes. Later in the afternoon four aircraft flew far afield, dropping their derision over suitable airfields up and down the country. The reactions varied considerably. Valley and Chivenor made impolite but friendly gestures by telephone, but the Lightning bases of Fighter Command were very upset. Wattisham demanded that the Squadron send someone down to pick up the paper, and Coltishall was sufficiently moved to send two Lightnings on a retaliatory mission. Unfortunately, their leaflets scored a direct hit on 54's hangar.

Five aircraft, led by the CO, flew down for a nostalgic visit to No 1's spiritual home at Tangmere on 4 April. After completing anniversary flypasts at Brighton and Chichester, they were royally entertained by the locals before returning to West Raynham on the following morning — the last fighters to fly from Tangmere.

On the way back to base, Flt Lt Alan Pollock broke away from the formation. He flew down the Thames at low level and beat up the Houses of Parliament; on his way out he continued down river, rocking his wings in salute at the RAF memorial and the statue of Lord Trenchard on the embankment. He then made headlines in the world's press by flying under the top span of Tower Bridge as an afterthought.

He was placed under arrest shortly after landing, to await the inevitable proceedings. A few days after the incident a group of MPs tried to table an all-party motion to secure his release. Before this, or any other action could be taken, Al was taken ill with pneumonia and was subsequently discharged from the RAF on medical grounds with a full pension.

Alan Pollock's place as 'B' Flight Commander was taken by Flt Lt Ken Parry, and the Squadron left for El Adem on 24 April. Before undertaking the usual APC, the Squadron played a leading part in Exercise 'Crayon', in which simulated strikes were flown against armoured forces manoeuvring in the desert. In this the pilots were joined by the Buccaneers of 801 Squadron of the Fleet Air Arm — the Navy's No 1.

The Squadron contributed eight Hunters as part of No 38 Group's offensive support display at the Queen's Review of the RAF at Abingdon on 14 June. No 1's show consisted of a simulated fighter sweep followed by two dummy attacks in support of an airborne landing from Argosies.

In August it was confirmed that No 1 would be 'going vertical' again during 1969, as the first squadron in the world to be equipped with the Hawker Siddeley Harrier. The news caused considerable excitement, particularly when it was heard that most of the existing pilots of the Squadron would move to Wittering to convert en bloc to the new aircraft.

Sqn Ldr Spike Jones came to the end of his tour on 20 September, having led the Squadron through two eventful years with skill and understanding. When dined out by his officers, Spike Jones was given an outsize silver medal that had been cast with his own profile on the face and the Squadron emblem on the reverse. He was promoted to Wing Commander and posted to command No 74 Squadron in Singapore, but a spinal operation prevented him from taking up his post, and he was transferred instead to the British Liaison Staff in Washington.

Spike Jones was replaced as CO by Sqn Ldr Bryan Baker who, by the end of the year, had led No 1 on a number of interesting exercises. One, an ultra-low flying trial in Northern Ireland, was notable for the Squadron's contribution to road safety in Ulster. Four police constables had the misfortune to meet Fg Off Paul Wilcox coming over the brow of a hill on the wrong side of the road, but what really put them off was that Paul was driving a Hunter at 450kt and 50ft. Understandably this caused them to leave the road, fortunately without injury.

The last four Hunter pilots to arrive on No 1 were posted to West Raynham at the beginning of 1969. These were Fg Offs Chris Humphrey and Neal Wharton from No 20 Squadron in Singapore, and Fg Off John Grogan and the author, who had just returned from Hunter squadrons in Bahrain.

The first half of the year was hectic. In March we took the Hunters to Skrydstrup in Denmark, and to Gutersloh in Germany for various exercises. To a newcomer the versatility of No 1's programme was very impressive. One week the Squadron was co-operating with Danish Forward Air Controllers in Schleswig-Holstein; then, after a few days of combat and simulated strikes from West Raynham, we were flying visual reconnaissance sorties in Germany, followed by a firepower demonstration for the Army on Sennelager range. Next, off to Leuchars to learn how to shoot down a helicopter, before returning to base to prepare for the Squadron's final trip to Libya.

With the advent of the Harrier pressing upon us, we welcomed a new Commanding Officer just before our departure for the last APC at El Adem. Wg Cdr J. A. Mansell took command on 10 April. He had commanded a Flight on No 1 during the Suez campaign and was the first Wing Commander to command the Squadron. Meanwhile, Bryan Baker was so keen to fly the Harrier that he asked to stay on as 'A' Flight Commander. Ken Parry was posted out to No 8 Squadron in Bahrain and Mike Webb retained command of 'B' Flight for the remaining months of the Hunter.

Jimmy Mansell was killed on 21 May, shortly after our return from EL Adem. He was getting the upper hand of a low level combat with John Thorpe at the time, in hazy conditions off the Norfolk coast, when John lost sight of him; eye witnesses near Wells saw a Hunter dive into the sea and explode. His loss came as a tremendous blow to us all, particularly because in the tragically short weeks of his tour he had taken a firm and kindly grip on the Squadron. The loss to his family too was complete; both his father and brother had been killed while serving in the RAF. Bryan Baker resumed acting command while a new CO was sought.

The Squadron's final regular detachment to Gibraltar took place in June, and I flew one of the final pair of Hunters down with Flt Lt Cid Sowler. It fell to us to bid farewell to the Rock on behalf of No 1. In company with Tim Smith, Neal Wharton and 20 groundcrew, I remember, rather hazily, that we achieved the aim pretty thoroughly.

The Squadron's worthiest farewell to the Hunter took place in Denmark in June. A team from the West Raynham Wing had been invited to Skrydstrup to take part in the annual 'Bullseye' Fighter Weapons Competition. This was run by NATO's North Europe Command, which meant that the team would be appearing as guest artists. The Danes, Norwegians and Germans had been practising for the shoot since early in the year, whereas the Raynham team was only able to train for a fortnight and had to modify the Hunter's attack profiles to suit the competition. Mike Webb, John Feesey and Tim Smith represented No 1, and the team was led by Wg Cdr Ron Wood, OC Flying at Raynham. The Standard and FAC phases of the competition were won without difficulty with the help of Flt Lt Pete Maillard, the team's Forward Air Controller, despite stiff competition from the F-5s and F-104s which had the benefit of much more modern sighting equipment.

On this triumphant note we prepared to leave West Raynham. Several of the pilots would not be coming with us. Mike Webb was promoted to Squadron Leader and posted to Bahrain, while Capt Pete Albrecht went home to the States to refresh on the F-105 before returning to Vietnam for a third tour of duty. In their place we gained Sqn Ldr 'Porky' Munro, who had flown on the Kestrel Tripartite squadron, and Capt Arthur Hall of the USAF.

When we left West Raynham we also left our Hunter groundcrew behind: a new set of white-coated Harrier technicians would be awaiting our arrival at Wittering. The hard work and cheerful spirit of the Hunter men had been an inspiration to every pilot on the Squadron, and when we said goodbye to them at Raynham we felt that they symbolised the last of a dying breed in the RAF. We were wrong, of course, because their successors soon proved to be just as resourceful and cunning. But we, the pilots, so soon to be thrust into the cold age of modern technology, were very sorry to be losing them.

CHAPTER SEVENTEEN

Full Circle

The pilots of No 1 Squadron moved house from West Raynham to Wittering at the end of June 1969. Before starting to fly the Harrier we were sent away for a short helicopter course at Ternhill to master the strange arts of hovering and flying backwards. After five sorties we found ourselves landing in tiny clearings with apparent ease under the wary eye of our instructors.

On returning to Wittering we completed another short course at the Aviation Medicine Training Centre at nearby North Luffenham. Here we picked up our new flying kit, which was distressingly complicated after the era of the Mae West and the Mk 1 Bonedome. Encased in our new skeletal harnesses, we suffered the indignity of being towed backwards across the airfield behind a lorry to practice our escape drills. Bud Hall, our USAF exchange oficer, was severely interrogated by his wife on returning home, because his underpants were covered in grass stains.

The only pilots qualified to fly the Harrier at Wittering at the time were the four instructors of the Harrier Conversion Team, who had just finished their own training under Duncan Simpson, the Chief Test Pilot of Hawker Siddeley. Led by Sqn Ldr Dick Le Brocq, the

The second vertical era begins: the Squadron's Harrier GR1s at Wittering in 1970, as a commemoration of the original tail-to-tail photograph at Clairmarais in 1918. The officers are, from the left: Bud Hall, Steve Jennings, John Grogan, Neal Wharton, John Feesey, Joe Sim, Stu Penny, Tim Smith, Bryan Baker, Maj Harry Jones (GLO), John Thorpe, Wg Cdr Ken Hayr, John Fowler, Mike Shaw, Brian Robson, Porky Munro, Roger Searle, W/O Matthews, Chas Cairns, Chris Humphrey, John Coates. *BAe*

team comprised Flt Lts Pete Dodworth, Ritchie Profit and Bruce Latton. All seemed to be jovially confident about the aeroplane, which made us very apprehensive.

As yet no two-seat Harrier or flight simulator had been delivered: the prospects of having to teach oneself to fly this powerful and revolutionary aeroplane were both exciting and daunting.

'A' Flight, led by Sqn Ldr 'Porky' Munro, was the first to start conversion, and received a week of ground school lectures at the end of July. After learning about the aircraft's systems and some of the flying techniques, the pilots practiced their cockpit drills and went out for their first taxying sorties.

Having mastered the nosewheel steering, which the Hunter did not have, each pilot lined up on the runway and made several full power runs to 100kt or so. This was to accustom us to the Harrier's shattering acceleration, and to teach us to swivel the exhaust nozzles and to use engine power to stop the aircraft after a conventional landing.

Porky Munro and Cid Sowler both flew the Harrier on the last day of July. The first sorties were flown, using a conventional take-off and landing, in company with a Hunter flown by one of the Harrier Conversion Team instructors. The main problem on take-off was to raise the undercarriage quickly before the speed limit was reached at 250kt, as one was still suffering from the tunnel vision caused by the unaccustomed acceleration. Once in the air the aircraft was a delight to fly from the very start, although we all commented on the lack of rearward vision in comparison with the Hunter. The conventional landing was terrifying. We crossed the hedge at about 165kt, and flew the jet straight on to the ground without rounding out, which ran contrary to all our experience and required tremendous self-discipline. Power nozzle braking was then used to slow the aircraft because the lightweight wheelbrakes could not be used above 50kt. We later discovered conventional landings to be much easier if the nozzles were lowered by 10°.

Wittering's airfield was not yet fully ready for Harrier training, so we returned to West Raynham to do our first hovers on the large concrete readiness platform at the end of the runway. The eight pilots of 'A' Flight were accompanied by the Station Commander of Wittering, Gp Capt P. G. K. Williamson. The Group Captain was already a tremendous Harrier enthusiast, having just commanded the RAF team in the *Daily Mail* Transatlantic Air Race, in which two Harriers had captured the imagination of New York and London with their vertical performances.

Each pilot's second sortie in the Harrier was conducted vertically. The instructor stood nearby with a mobile radio, while we taxied nervously out to the concrete pan. This sortie was amazingly easy. The nozzles were lowered to the hovering position and full power lifted the aircraft slowly and cleanly into the hover at 50ft. The instructor shouted encouraging comments from the sideline, and a small reduction of power brought the Harrier to earth in roughly the same place. Two more 'press-ups' then enabled the pilot to get the feel of the controls in the hover and to move gently around the pan, which often sent the ghoulish onlookers scurrying for safety.

The remaining conversion sorties taught us to accelerate from the hover and to slow down from forward flight. The latter exercise felt most unnatural at first, with the airspeed sinking below stalling speed as the Harrier became supported solely by four columns of hot air. The sinking feeling became only too real for Cid Sowler on one early sortie when he sank into a field of sprouts just short of the runway and the aircraft turned over. Fortunately it did not catch fire and Cid got out unscathed. That Harrier was flying again some months later, but Cid wisely decided that the Harrier was not for him. After a further incident he was posted to Chivenor to pass on his considerable skills as a fighter pilot to the next generation.

Wg Cdr Duncan Allison arrived at the beginning of August to take command of No 1 for his third tour on the Squadron. But like Sowler he found that the peculiarities of the Harrier do not always suit even the best of pilots, and he manfully stepped down. For the second time in a year Sqn Ldr Bryan Baker resumed temporary command.

On 14 October everything ground to a halt after Fg Off Steve Jennings aborted a take-off with an engine fire. One of the huge compressor blades had parted company from the engine and the fleet was grounded for two months while Rolls-Royce sorted out the problem. Meanwhile we kept ourselves amused by flying Jet Provost Mk 3s up at Cranwell.

By the end of 1969 we were flying the Harriers once more, and Wg Cdr Ken Hayr AFC arrived from the Phantom force at Coningsby to take command at the tender age of 34. We soon realised that the new CO was not a man to trifle with. His leadership and competence in

Wg Cdr Kenneth Hayr, Officer Commanding No 1 (Fighter) Squadron from January 1970 to December 1971. *Squadron Archives*

the air soon provided us with the inspiration that was needed as we settled down to the very hard years of evolving and proving the Harrier techniques that have since stood the tests of time and battle.

The work-up was now running well behind schedule because of the problems with the engines, so we were sent to Cyprus for a short detachment in March 1970 to avoid further delays caused by the bad winter. We had been joined by two new pilots, Flt Lt Stuart Penny and Fg Off Roger Searle, and we set off across France with an average of 25 Harrier hours per pilot.

We refuelled en route at Istres, Malta and Souda Bay in Crete, and large crowds gathered at each place to watch the aircraft arrive. At Akrotiri we were welcomed by the Lightning pilots of No 56 Squadron, who seemed envious of the attention-grabbing characteristics of a hovering Harrier. After a busy week in which we flew over 100 sorties, Ken Hayr led the first 'diamond nine' formation of Harriers, after which all nine aircraft slowed to the hover along Akrotiri's runway, turned into line and bowed to the crowd.

The new groundcrew really began to work as a team in Cyprus. They had come from a variety of non-fighter backgrounds and many were straight out of training. It had been difficult during the early months at Wittering to engender much Squadron spirit, but on that first overseas deployment we all began to pull together as a team.

The summer of 1970 was spent in many short exercises designed to prepare the Squadron to be able to operate away from a main airbase. We practiced rolling vertical landings and short take-offs from a number of farmers' fields around Wittering, and John Grogan was given the job of developing our camouflage techniques. HRH Prince Philip was one of many distinguished people who came to see us hiding in the woods and slithering our aircraft around the muddy fields of Wakerly Oaks, Twinwoods Farm and the like.

The groundcrew and the Squadron headquarters were also preparing to become mobile. The ground equipment began to be painted green, khaki working dress was issued, and by June we had completed two-day detachments with four aircraft at the disused airfields at Wymeswold and Bruntingthorpe. By now most of the aircraft had been fitted with the Ferranti Inertial Navigation and Attack System (INAS), complete with moving map and head-up display, which greatly increased the cockpit workload at first and was not without its teething troubles.

When the second Harrier squadron, No 4, was formed at Wildenrath in June, four of No 1's pilots (Bob Fowler, Chris Humphrey, Tim Smith and John Thorpe) were posted out to provide it with a nucleus of experience.

We returned to Akrotiri at the end of July for our first Armament Practice Camp. Initially using fixed sighting techniques, we soon began to achieve good scores with cannon, SNEB and retard bombs. But when we tried to use the new INAS to aim the weapons the results were terrible. Most pilots were inclined to ignore the 'kit' and rely on their innate skills to get high scores, but Wg Cdr Hayr rightly insisted that we persevere with INAS weaponry so that the technicians could get it right.

At the beginning of September 1970, No 1 Squadron was once more declared fully operational. This achievement was promptly demonstrated to the taxpayer at the Farnborough Air Display, where the CO, Porky Munro, Bud Hall and Stuart Penny thrilled the

The author landing at 50kt on the grass strip at Peterborough Flying Club, Sibson, in the summer of 1970. *AVM K. W. Hayr*

To the woods! Two of No 1's Harriers dispersed in camouflaged hides on the edge of a farmer's paddock near Wittering. It did not take long for the pilots and technicians to learn to make themselves comfortable and invisible under canvas.
AVM K. W. Hayr

crowd with a combination of high-speed passes and co-ordinated hovering among the woods on Laffan's Plain, where the Squadron had last performed vertical take-offs in airships in 1913.

One week later Wg Cdr Hayr led six aircraft north of the Arctic Circle to Bardufoss to exercise the Squadron's primary task in support of the NATO ACE Mobile Force. The aircraft lived in a hangar dug into a mountain, the groundcrew lived in an enormous heated bubble, while the officers slept in some awful Norwegian tents whose centre pole was the chimney of a wood-burning stove. The Boss's tent kept filling up with smoke,and it can now be admitted that this was not unconnected with some artificially damp wood and a clod of earth in the chimney. Ken Hayr got his revenge later by sending the author on a combat survival course with the SAS!

In October we deployed to Ouston in Northumberland, using a camouflaged site on the airfield, and this time we took our armament. The weapons were flown in by Hercules and were deployed in the woods with the aircraft. We flew armed reconnaissance and FAC sorties on Otterburn range, and by now we were all beginning to master the more basic arts of trying to live comfortably in tents. The exercise was enlivened by Neal Wharton's ejection on final approach following a fuel pump failure. He landed close to the blazing wreckage in a 40kt wind but was fit enough to discharge himself from hospital that night and return to the camp for a celebratory beer.

'Warty's' ejection cost us the first Harrier that had been lost in RAF service. There had been several close shaves; Porky had dug up a potato field during a miscalculated take-off, Joe Sim while hovering had wrecked an aluminium pad which then tried to wrap itself round the aircraft, and John Feesey crashed on take-off after hitting a seagull. The author carelessly dented two aircraft in two days, when trying to land in a clearing. But we were

Above:
En route to Cyprus, five aircraft gave a short display to the Italian Air Force at Rome. Looking suitably self-conscious, the pilots pause for a publicity photo during a tour of the Coliseum. From left to right: Wg Cdr Ken Hayr, Sqn Ldr Porky Munro, and Flt Lts John Feesey, Neal Wharton and Joe Sim.
AVM K. W. Hayr

Below:
Everybody wanted to come and see the Harriers. Here Wg Cdr Hayr shows Lord Carrington around the new aircraft. This one is standing on alloy MEXE-planking, from which the vertical landing pads are constructed. *AVM K. W. Hayr*

On board HMS *Ark Royal* in June 1971. This was the first deployment afloat by operational V/STOL jets. The Admiralty watched with interest, and soon confirmed its decision to buy the Sea Harrier.
D. Hart

incredibly lucky to have survived that first innovative year without losing an airframe or, worse still, a pilot.

After further weapon training at the new NATO range at Decimomannu in Sardinia, we were ready for our first full deployment away from an airfield. In March 1971 we flew into the disused aerodrome at Milltown near Lossiemouth for Exercise 'Spadefoot'. We fired full pods of SNEB at Cape Wrath and dropped 1,000lb retard bombs on Tain range. Brook Blackford caused quite a stir by returning to Milltown with a clean aeroplane, having dumped his tanks and SNEB pods 'somewhere in Scotland'. The press were visiting at the time, and the world was treated to *Daily Mirror* headlines such as 'Rocket Peril in the Hills!' The pods were eventually found on the clifftop at Cape Wrath range.

In June the Squadron gained further fame at the Paris Air Show, with a five-man team led by Wg Cdr Hayr. Their display drew rave reports from the world's aviation press despite an incident in practice when Porky Munro's outrigger collapsed on landing. Our new USAF exchange officer, Louis Distelzweig, was in the control tower as duty pilot, and he was happy to reassure the Boss that the show had only just opened to the public so 'only about 40,000 people saw the Goddam prang!'.

The triumphant team returned to Wittering, reeking of garlic and gallantry, in time to despatch four aircraft to HMS *Ark Royal* off Rosyth. The CO performed the Squadron's first deck landing with great panache. He chose to do a rolling vertical landing, but skidded on the wet deck and almost collided with the assembled Phantoms and Buccaneers up at the sharp end. He was followed by Sqn Ldr Dave Willis, Joe Sim and Neal Wharton, who executed rather more sedate vertical landings. This first carrier deployment by operational V/STOL aircraft finally convinced the Royal Navy that the Sea Harrier and the 'through-deck cruiser' were worth buying. On one occasion *Ark Royal*'s own aircraft were diverted ashore in bad weather. Wg Cdr Hayr persuaded Capt Roberts to allow Dave Willis, who was also airborne, to try to approach the carrier, which he did without difficulty, and the naval pilots were very impressed with the versatility of the Harrier. The ship's company was also entertained to some accurate gunnery against a towed splash target. The INAS was still playing up, however, and proper alignments were difficult to achieve on a moving ship, even when using the special FINRAE equipment. The dive bombing display was thus less accurate than the strafing, and Joe Sim was wryly credited with a direct hit on the sea.

In July the first Squadron reunion and open day for some years was held, and many former members arrived by air in aircraft types flown by No 1 in the past, including a Hunter 9, a Meteor 8 and a Spitfire. Joining them on the tarmac was the Squadron's Spitfire 21, now lovingly restored after a road crash en route from West Raynham. We were proud that day to welcome a spritely Air Cdre Philip Fullard as our senior and most distinguished guest.

Then, on 3 August, we lost Louis Distelzweig. He was leading a four-aircraft strike sortie to Wales when his nozzles jammed on take-off. After some inspired flying he was preparing to land when the nozzles ran fully forward and the aircraft stopped dead in mid-air. Louis left his ejection too late and was killed, a tragic end for a delightful American who had won the Silver Star for strafing trains at night in a F-100 in North Vietnam. A grieving Squadron was grounded for several weeks while the engineers eradicated any possibility of a recurrence.

Our previous foray to Bardufoss was good practice for a full winter exercise at Gardermoen near Oslo in December, in which the full Squadron was dug into a wood under several feet of snow, together with a film crew from HTV. Our arrival was dramatised by Steve Jennings, whose nosewheel steering froze solid and flipped him into a snowbank upside down. Steve got out with a shaking but the Harrier had to go home in a Belfast. We had been equipped with some excellent winter equipment, including a hot shower tent into which Neal Wharton promptly disappeared with a girl from the TV crew.

On returning from Norway, No 1 greeted the arrival of 1972 with a phase of flight refuelling from Victor tankers in preparation for deployments further afield. My own final sortie on No 1 was a marvellous scenic tour of Scotland and Northumberland that lasted 3½ hours, pausing twice to take on fuel from a friendly Victor.

It was time for a number of farewells. Wg Cdr Hayr went to Staff College and was succeeded by Wg Cdr Eric Smith. Fg Off Chas Cairns, our ever-efficient Adjutant, handed the books over to Plt Off Makin, and I was posted to No 3 Squadron at Wildenrath. New pilots arriving included John Lloyd from helicopters and Keith Grumbley from No 8 Squadron in Bahrain. Dave Linney and Kit Adams became the Squadron's first first-tour Harrier pilots. Shortly afterwards Sqn Ldr Richard Kyle replaced Derek Bridson as Senior Engineering Officer.

In March the Squadron deployed to Kinloss for Exercise 'Snowy Owl', and for the first time

Flt Lt Joe Sim slows to the hover near *Ark Royal*, as another No 1 Squadron Harrier lines up for take-off. The Fleet Air Arm's pilots readily agreed that it was easier to stop and then land on a deck, than to land on a deck and then stop. *D. Hart*

two dispersed sites were used simultaneously when six aircraft and crews moved to a sub-site at Milltown.

By now the Squadron had practised every major aspect of deployment that would be required in wartime, and the pilots had become skilful with all the weapons in the Harrier's inventory. This was in spite of losing some more of the most experienced pilots, for Joe Sim, Steve Jennings and John Grogan were posted out during the summer. The new pilots were able to gain experience more quickly because the flight simulator was now in action at Wittering; the Squadron also received its two-seat Harrier T2. None of this prevented Kit Adams from attempting a hitherto untried inverted hover on May Day. He ejected close to the control tower and suffered minor facial injuries.

After further training with the Victor tankers, the Squadron deployed to Cyprus in May for Exercise 'Bold Front'. Meanwhile Flt Lts Neal Wharton and Dave Fisher had been chosen for a Harrier display at the Toronto Air Show, which gave the Squadron the opportunity of its first trans-Atlantic flight, accomplished in 8 hours 24 minutes for the 3,127-mile flight. The Harrier display was the climax of the Toronto Air Show and the whole team returned to Wittering having been thoroughly spoilt in Canada.

The annual NATO autumn Exercise 'Strong Express' took the Squadron to Tromso in northern Norway for the first time. Tromso was to become a regular deployment base in the future for No 1's role within the ACE Mobile Force. On this occasion it was billeted at the quaintly named Sydspissen Permcentre, and found that the social life of Tromso was distinctly promising.

In recent months Flt Lt Roger Searle had been posted, to be replaced by Flt Lt John Gamble, and Sqn Ldr Jack Pugh arrived to take over 'B' Flight from Sqn Ldr Dave Willis, who later went to HQ No 38 Group to look after the interests of the Squadron as the Harrier staff officer. He was awarded an AFC in the next list for his enthusiastic leadership. Capt Tom Plank filled the vacant USAF exchange post, having recently flown 173 combat missions in Vietnam.

In November the annual APC was held in Decimomannu, when the Scottish heirarchy of the Squadron celebrated St Andrews day with a vengeance. The jazz band was now flourishing under Dave Fisher and it was joined on this occasion by Sqn Ldr Jock Beaton on bagpipes and the resident RAF staff officer, Sqn Ldr Alec Wedderburn, on haggis.

No 1 returned to Milltown in February 1973 for Exercise 'Sky Mist' during which Jack Pugh was awarded the AFC for saving his aircraft after a bird strike in the Cairngorms. The engine had surged and lost power, but he had managed to relight it while sinking over a ridge at 200kt.

After a weapons camp in Cyprus in May, at which the Hilly Brown trophy was won by Flt Lt Keith Grumbley from Kenya, the Squadron enjoyed a NATO squadron exchange with Grupo 103° of the Italian Air Force at Treviso. In July three aircraft, flown by John Marshall, Dave Fisher and Dave Linney, embarked on HMS *Bulwark* for a short exercise afloat.

In winter plumage, two Harriers taxi out of their hides at Gardermoen, near Oslo, during Exercise 'White Tent' in December 1971. The whole Squadron spent 10 days camping in this wood, and operated successfully in temperatures down to −36°C.
AVM K. W. Hayr

The NATO autumn exercise of 1973 was called 'Absalom Express', and this time No 1 was assigned to Vandel in central Denmark, where all the support equipment had to be flown in, and the Squadron was required to practise with gas masks and other NBC equipment in the field. Then, after the usual winter APC in Sardinia, Wg Cdr Eric Smith was able to hand over his command of No 1 Squadron to Wg Cdr P. P. W. Taylor in the spring of 1974, confident that his unit had made great progress in the past two years and that it was now fully capable of operating away from base in wartime.

In June the Squadron was required to demonstrate its ability to sustain a very high sortie rate for a three-day 'war'. Exercise 'Big Tee' took place at Watton in Norfolk and the Squadron strength was beefed up by some pilots and airframes from RAF Germany to make sure that 12 aircraft were constantly available. In three days 364 sorties were flown, simulating attacks on troops and vehicles in the nearby Stanford battle area. Despite the artificiality of the exercise, a lot was learned about rapid tasking of sorties, especially when the pilot remained in the cockpit during turn-round.

At the SBAC Display at Farnborough in September, Wg Cdr Taylor was required to produce a team of 10 aircraft to execute a complicated sequence that included formation hovers, high-speed runs and simulated attacks. There had been little time available to practise in the previous weeks and Sqn Ldr-Dave Lott recalled that the first rehearsal was a shambles: 'The weather was appalling, and the Boss took the 4-ship in close formation over the President's tent very low. Gp Capt George Black, who was in charge of the RAF team at the show, was very cross at the debrief!' However, their morale was boosted by Air Cdre Peter Latham, who gave them a very encouraging talk that evening. The show went perfectly on the day, but they were saddened by the crash of the Blackhawk attack helicopter, which happened very close to them as they taxied out. Furthermore, Tom Plank, who led the high-speed low pass, was filmed at 35ft while doing over 500kt, which resulted in further apoplexy in high places.

After the winter APC in Sardinia and Exercise 'Cold Winter' at Tromso, there was a most enjoyable NATO exchange visit with No 311 (Boelcke) Squadron of the German air force, which flew F-104 Starfighters. The usual inter-squadron banter developed into a challenge from the Germans that a Harrier could not beat a Starfighter to 10,000ft from a standing start. After much midnight engine tweaking, the race took place between the two-seat trainers of each unit. Jack Pugh rode in the F-104 to see fair play, and was alarmed to see that the fuel gauges had been taped up in the rear cockpit. Flt Lt Pete Harris flew the

Saying goodbye to the American exchange officer. Capt Tom Plank is 'towed past' in the Squadron's Spitfire 21 before his return to the USAF. Since World War 2, exchange officers from America, Holland and Pakistan have served with No 1.
Flt Lt H. de Courcier

T4 Harrier with a German umpire, and he beat the Starfighter to 10,000ft by seven seconds. The Harrier was carrying full fuel and went on to complete an hour-long combat sortie, whereas the F-104 darted straight home to land with empty tanks.

On the Squadron's return from a further APC at Decimomannu in October 1975, it was placed on one week's notice to deploy to Belize, in Central America. The former colony of British Honduras was under threat of invasion from Guatemala and the Harrier was the only British fighter that could operate from the short single runway at Belize Airport.

After some frantic preparation which included discovering the whereabouts of Belize, six Harriers deployed to the Caribbean for Operation 'Nucha'. Refuelling in flight from Victor tankers from Marham, they crossed the Atlantic to Goose Bay in six hours, and on the following day they staged through Bermuda before the final four-hour leg across Cuba into Belize. The political situation was very odd — the Americans were supporters of the right-wing regime in Guatemala, while Fidel Castro had announced his support for the new government of Belize — and the Squadron was therefore unable to use American bases during the deployment.

The first two aircraft to arrive at Belize were flown by Sqn Ldr Dave Lott and Flt Lt John Buckler. The advance party of groundcrew performed an operational turn-round and replaced the big ferry tanks with 100gal combat tanks. Within 20 minutes they were airborne again, to 'flagwave' Belize City and to fly along the border, letting both sides know that Britain had reinforced the tiny Belize garrison. The effect was startling. The Guatemalans did not invade and the Belizeans danced in the streets in celebration of their deliverance. Once the Squadron had settled in at Airport Camp, they found themselves to be celebrities with the local population who took to naming their children after individual pilots or just simply 'Harrier'.

Two of the Squadron's Harriers on combat air patrol over Belize City. No 1 Squadron has deployed twice to Belize, using in-flight refuelling, in response to threatened invasions from Guatemala.
Flt Lt H. de Courcier

The living conditions at Airport Camp were dreadful. The groundcrew camped with the aircraft in six dispersed hides clustered around the end of the runway; it was hot and humid, and they had to contend with snakes, huge spiders and poisonous centipedes. The pilots' accommodation was little better; they shared tiny compartments in a huge airless nissen hut, and fed in shifts in a mess that was overcrowded by a factor of five.

The flying turned out to be excellent. The regular border patrols took the Harriers

Above:
Wg Cdr Peter Taylor with his officers and senior NCOs in Belize in January 1975. Sitting either side of him are Flt Lt Peter 'Bomber' Harris and Sqn Ldr Dave Lott. *Flt Lt. H. de Courcier*

Below:
An engine change in the field calls for ingenuity. Here a Belizean crane is pressed into service.
Flt Lt H. de Courcier

through spectacular mountains covered in primeval jungle, and they were scrambled several times to intercept suspicious aircraft: these were mostly DC-3s and DC-6s unaware of the tension on the border, and which had not bothered to file a flight plan or whose business was less than legal. Low flying among the coastal mangroves and cays was made hazardous by large birds, as Flt Lt Bernie Scott found to his cost. He hit an eagle and was forced to eject into the sea, destroying the author's old XV788 in the process.

As Christmas approached, the political situation eased. Some of the detachment were able to fly home and those who remained were royally entertained by Belizean society, both in the city and on the delightful offshore cays where many of the wealthy had weekend houses. Flt Lt Henry de Courcier found his future wife in Belize, and the Squadron formed many enduring friendships.

Just before Christmas, Guatemala city was devastated by an earthquake which led the government there to set aside any immediate territorial claim to Belize. The Squadron was therefore released from Operation 'Nucha' and returned to Wittering, but not before the Prime Minister of Belize had asked Dave Lott, the detachment commander, if two aircraft could be provided to beat up his city at low level. 'You don't often get an opportunity like that!' said Lotty. 'I went off with Pete Harris, and we roasted around the rooftops for a while, before landing to take part in the farewell celebrations.' The aircraft were dismantled and came home unceremoniously by Hercules. But they had not seen the last of Belize.

It took No 1 a long time to get back to full efficiency after 'Nucha'. Wg Cdr Taylor spent his final months in command organising the intensive training and maintenance that was needed before the Squadron resumed its place in NATO's front line. At Decimomannu in July 1976 the pilots re-sharpened their skill with weaponry, and Flt Lt Peter Bennett won the coveted Hilly Brown Trophy. Shortly afterwards Peter Taylor handed over the Squadron to Wg Cdr Jeremy Saye.

During the autumn of 1976 it deployed to Oerland, on the west coast of Norway, for Exercise 'Teamwork', in which the Squadron was joined by the US Marine Corps. The Marines were flying their own AV-8B Harriers from USS *John F. Kennedy*, but they later came ashore to try their skills at camping. Dave Lott remembered:

'They didn't have a clue. They got all their aircraft and vehicles bogged in the mud and then got terribly over-excited. Their Admiral went mad when he learned that No 1 had flown more sorties with 12 Harriers than the whole Marine Air Group could manage with over 80 jets, including those on the carrier.'

Another exchange visit took place in November, this time with the French 'Cigognes' Mirage squadron at Dijon. In addition to a number of memorable nights spent in wine cellars in Nuits St Georges, the pilots also enjoyed some good flying, which included some composite attack sorties in formation with Mirages and Jaguars. They came home very impressed with the quality of the French Air Force. On their last night in France the two-seater was stolen by the French groundcrew and was resprayed pink and purple. Gloss paint had been used, so the T-bird returned to Wittering looking very embarrassed.

1977 opened with a variety of training exercises which culminated in a Tactical Evaluation (Taceval), using RAF Leeming to simulate a bare airfield in Turkey. On returning to Wittering, intelligence reports began to predict a further invasion of Belize from Guatemala, and No 1 was placed on readiness at the end of June. This time it had sufficient warning to carry out an intensive phase of air combat training and the pilots fired their guns for the first time against a towed target. The INAS had never been properly tuned for air gunnery and the results were poor enough for the pilots to resort to fixed sighting.

The advance party of Sqn Ldr Lott and Flt Lts Scott and Watts arrived in Belize by VC-10, 24 hours before the first pair of Harriers. Lott arrived to find the Belizeans in a state of panic about the invasion, yet the resident Army battalion was still playing sport, with its weapons firmly locked up in the armoury. The SAS had also arrived recently, and its Major was so appalled that he was preparing to take his men off into the jungle to await the Guatemalans. The pillow tanks, for holding fuel at each hide, had not yet arrived, and most of the airport's bowsers were unserviceable. After frantic preparations with nets, weapons and spares, all was ready for the five Harriers when they arrived, led by Wg Cdr Saye and Sqn Ldrs Pete Day and John Finlayson. At dusk, Wg Cdr Derek Hine, the Air Commander, received a report from a private pilot that the A-37s at 'Guat City' were bombing up, clearly in preparation for a dawn

One of six No 1 Squadron Harriers sits on armed alert in its hide at Belize airport, in July 1977. The Guatemalan threat included a squadron of A-37 Dragonflies which were believed to be armed with Sidewinders. *Flt Lt H. de Courcier*

raid on Belize Airport. Jerry Saye exhausted after the long ferry flight, sat down with his Flight commanders to work out their response. Three aircraft were to get airborne before dawn, leaving two on alert. It was known that the A-37s would not have the range to return to their base, and would have to refuel from jerry-cans at a coastal airstrip just south of Belize. Those that survived the Squadron's air patrol would be destroyed on the ground, and the war would be over by lunchtime!

In the event, the Guatemalans never came. Their foreign minister was issuing threats at the UN when Ted Rowlands MP was able to inform him of the British reinforcement. The political pressure stayed high for about three days, but then the Guatemalan troops were pulled back from the border and the threat gradually receded.

No 1 Squadron began to dig in for a long stay. While the Harriers flew border patrols in contact with Army and SAS outposts, the groundcrew and mobility staff worked to improve the hides and the logistics area, with better camouflage and sandbagged walls. The tents were mounted on fresh gravel to keep down the insects and proper drainage ditches were dug. After the first month some porta-cabins were flown out, which doubled as flight offices and sleeping accommodation for the groundcrew.

One of the Squadron's civilian friends flew into 'Guat City' on business and returned with the news that some of the A-37s were armed with Sidewinder missiles. This led to an attempt to borrow some surplus Sidewinders from the Royal Navy and to modify the aircraft wiring so that they could be fired. This plan was vetoed, because the British Government was keen to cool the situation as fast as possible, but some useful feasibility work was done which would prove very useful five years later.

By the autumn of 1977 it was obvious that the Belize deployment would be long drawn-out, so the Squadron's establishment was increased to permit a proper rotation of personnel without too much damage to No 1's NATO commitments. Pilots and groundcrew from the RAF Germany squadrons were later fed into this rotation, and No 1 Squadron ceased to be solely responsible for the Belize detachment in June 1978. Just as the Squadron was getting together for the first time in a year, Wg Cdr Jerry Saye handed command of No 1 to Wg Cdr Dickie Duckett, a former leader of the Red Arrows, on 17 July.

Exercise 'Northern Wedding' took place in Orkney and Shetland, in a blaze of publicity in September. The Harriers were based at Kirkwall, and operated in support of the British and Dutch Marines. The ships and aircraft totalled 1,000 and there were 40,000 troops involved. The American system of close control meant that the Harriers were not permitted to fly below 800ft, but otherwise the exercise was a great success. The weather was often very wild, and Wg Cdr Duckett remembered looking out of his tent to see the latrines flying off into the night, trailing great streamers of paper.

Heavy snow at Wittering, and a freezing deployment at Barkston Heath, prepared the Squadron for an enjoyable exercise at Bardufoss, but it then warmed up in Sardinia in time to resume the Belize detachment in July 1979. During the autumn, unusual troop movements took place in Guatemala, and the A-37s were moved up to their forward strip at Puerto Barrios. No 1's detachment sat on eager alert for three days but the panic died down within a week.

On 21 September the CO and Chris Gowers collided during an air combat sortie to the east of Wittering, Duckett ejecting from a spin. Gowers tried to fly his crippled aircraft back to base, but a wing fell off and he too had to bale out. Sadly, his Harrier crashed at Wisbech, killing two people. Another aircraft was lost at Holbeach range in November when Ross Boyens picked up a ricochet that had probably been fired by Mike Beech, with whom he was sharing the range slot.

The first half of 1980 was a busy time for No 1. At Tromso for Exercise 'Anorak Express', it fought with American, Canadian, Dutch and Norwegian air defence fighters. The Squadron's aggressive tactics helped it to counter the faster opposition very successfully, while the whitewashed winter camouflage made the Harriers even more difficult to see against the Arctic terrain.

Back at base the work-up was in full swing for Exercise 'Maple Flag 5' in Canada, where 40 or 50 tactical aircraft could simulate wartime conditions very realistically. The Harriers crossed the Atlantic with their Victor tankers — a total of 13 hours via Goose Bay — without incident, and the pilots enjoyed the opportunity of unlimited low flying in the wide open spaces of Alberta. 'The exercise was characterised by the Americans,' said the CO, 'who kept crashing their jets into the dry forest and setting the range on fire, which was most tiresome. I think they lost three while we were there.'

Tactical Evaluation took place at Gutersloh in September, while the resident Harrier squadrons were deployed in the field for the Army Exercise 'Spear Point'. The Squadron operated from No 3 Squadron's hardened shelters, and Sqn Ldr Bob Iveson led a co-ordinated attack by 18 Harriers on a massed force of armour.

In October 1980 No 1 enjoyed a sunny NATO exchange at Montijo in Portugal as guest of a G-91 squadron. While there the pilots lived in a hotel in Lisbon and went to work each day by boat. During the same month they were proud to welcome HRH the Duchess of Gloucester to Wittering from her home nearby.

Wg Cdr Duckett came to the end of his tour in March 1981, just after the end of another enjoyable exercise in Tromso. Wg Cdr P. T. Squire AFC took command in time to welcome another VIP in the shape of former Vice-President Walter Mondale.

During the summer No 1 was fully committed on its usual round of exercises. The annual Taceval took place in two phases; the first part at Wittering tested its state of readiness and reaction to a surprise deployment, and the Squadron was awarded a coveted Grade 1 rating. Part II took place under canvas at Vandel in Denmark, where all demonstrated their ability to survive and operate under air and chemical weapons attacks.

Early in 1982 the Squadron began to prepare for another 'Maple Flag' deployment to Canada, which was scheduled for early April, and completed some of the work-up training before the usual spring deployment to Tromso.

No 1 returned from Tromso at the end of March in good heart. The training had been going well and the Squadron had a nice balance of experienced pilots and groundcrew who were able to work together as a well-drilled and friendly team. They were all looking forward to the trip to Canada and to a routine summer of good flying, hard work and even a spell of leave.

And if any of them had time to read a paper, they might have spotted a small item about some scrap merchants who had hoisted a flag on somewhere called South Georgia.

CHAPTER EIGHTEEN

Operation 'Corporate'

The news of the Argentine invasion of the Falkland Islands was received at Wittering with predictable interest. Wg Cdr Peter Squire realised that No 1 Squadron's unique ability to operate away from an airbase, or from an aircraft carrier, would make it a prime candidate for deployment. But when he and his pilots looked at the map they realised that the logistical problems facing any British Task Force would be almost insuperable.

They had reckoned without the dynamic reaction of the government, the Services and the British people, which enabled all peace-time constraints to be put aside. The leading ships of the Task Force sailed within days of the invasion and the operational staffs of all three services began working round the clock.

It was a stroke of fortune that the key RAF Staff Officer at MoD was a former CO of No 1 Squadron. AVM Ken Hayr, was Assistant Chief of the Air Staff (Operations), and his assistant was Wg Cdr Joe Sim. Both men had flown the Harrier from the start and had embarked aboard HMS *Ark Royal* during No 1's first carrier exercise in 1971. By 10 April the Squadron had been warned to prepare to deploy to the Task Force, initially to reinforce the Sea Harriers in the air defence role. At the same time the rest of the Harrier force was warned of the need to support No 1 with aircraft and, later, pilots.

The first task was to select and prepare the aircraft. Those approaching routine servicing were rejected and 16 'good goers' were selected. The groundcrew adopted a 24-hour shift system, supported by Wittering's Engineering Wing and British Aerospace, and they began to modify the chosen aircraft for deck operations. This involved the fitting of lashing shackles to the outriggers, drilling seawater drain-holes in suitable places, and enabling the nosewheel steering to be connected to the rudder bar when required.

The pilots had meanwhile flown the Atlantic to Canada! They had ferried eight RAF Germany Harriers, as planned, to Goose Bay for Exercise 'Maple Flag' because the RAFG pilots were not trained for flight refuelling. The CO was surprised that this deployment was not cancelled, but the pilots returned to Wittering by Nimrod on 14 April, correctly feeling that the eight-hour transatlantic hop had been a useful dummy run.

Amazing efforts were now being made in various quarters to provide the Squadron with the correct weapons and training for the task ahead. Many RAF officers — the author among them — had pleaded over the years for the Harrier to be equipped with the means of firing the AIM-9 Sidewinder missile, but the funds had never been available for the preparatory work to be done. In the event this was achieved by the engineers in under three weeks, using redundant wiring in the wings and by manufacturing the switchboxes at Wittering. Furthermore, an I-band transponder to enable the pilots to make carrier approaches in very bad weather was designed, manufactured and fitted before the aircraft flew south. Similar work was being started with a chaff and flare dispenser, bought off-the-shelf in America, which was fitted later to the inside of the rear equipment bay door.

One crucial effort was the production of FINRAE Mk 2, by Ferranti, which would enable the INAS to be aligned properly from a moving ship. Without INAS the pilots would be denied accurate weapon-aiming and navigation information, and they would have no head-up instruments for ultra low-level flying or for carrier recoveries in bad weather. Joe Sim got the ball rolling by asking Ferranti how quickly the company could resurrect FINRAE, which had not been tried since he himself had used it on *Ark Royal* in 1971. 'Oh, about six months . . . came the reply. 'No', said Joe, 'I meant VERY quickly!' A completely redesigned FINRAE was flown to join the Task Force at Ascension Island 18 days later, after a magnificent effort by all concerned.

The pilots were also working round the clock. Flying aircraft from No 233 OCU and RAF Germany, they began training hard for every eventuality. They started with an intense phase of air combat training against each other and then against Hunters, Lightnings and Phantoms from various 11 Group squadrons. These sorties, and the debriefs that followed them, were vital confidence-boosters for the men about to be launched into the unknown.

It was still not clear what role would be required of them in the South Atlantic, so they were also cleared, after some dithering at Strike Command, for 'operational low flying' at ultra low level and high speed. In the event the importance of this training cannot be underestimated, and more aircraft would probably have been lost without it.

On 13 April Sqn Ldrs Bob Iveson and Bruce Sobey (No 1's senior engineer) were sent to Liverpool to check out the container ship *Atlantic Conveyor* to see if it could be used to carry the Squadron southwards to join the fleet. With a strengthened take-off pad forward and rows of containers on each side to protect the aircraft from the sea, it would do. Capt North duly set sail with his ship, while a dockyard crew from Devonport modified the ship en route to Ascension Island, where the Harriers would embark.

By 19 April most of the pilots had been to the Sea Harrier base at Yeovilton for practice take-offs on the 7° ski-jump, which they all found to be remarkably easy. They were now joined at Wittering by five former No 1 Squadron pilots who had been hurriedly posted in to bolster the unit with their experience: Sqn Ldrs Tim Smith and Pete Harris, and Flt Lts Ross Boyens, Jim Arkell and Clive Loader. 'Bomber' Harris was a natural choice because he had carrier experience, and he flew with the Squadron regularly as part of his job with the Central Trials and Tactics Organisation. These five were soon followed by Flt Lt Jeff Glover, who had been away on a NATO Tactical Leadership Course at Jever.

During the following week they dropped cluster bombs on splash targets, and completed trials with Naval 2in rocket pods and they also practised attacking a Type 42

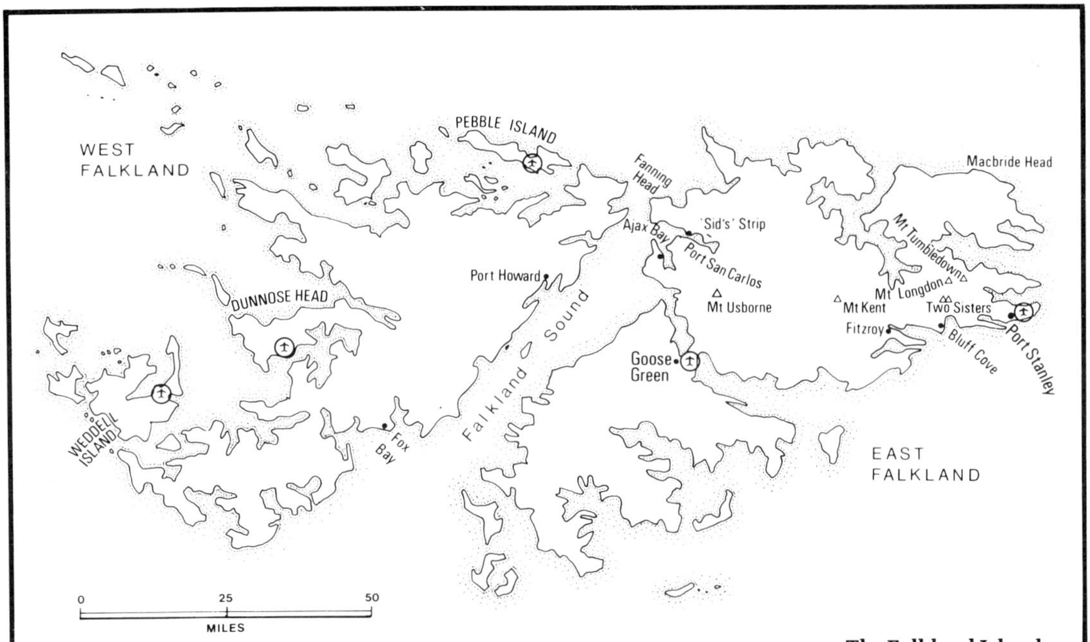

The Falkland Islands.

destroyer which was identical to those used by the Argentine Navy: this showed the pilots what responses to expect on the Radar Warning Receiver in the Harrier's cockpit.

On 29 April AIM-9G missiles arrived at Wittering, and the pilots began training with the modified aircraft, initially using dummy rounds to experience the Sidewinder's warning 'growl' in the earphones, which tells the pilot when the missile has locked on. Six pilots fired live missiles on 30 April against flare targets towed behind a pilotless Jindivik. Bob Iveson's 'Winder failed to guide, but the other five scored direct hits — a tribute to the training they had received from Capt Skip Beasley, the USAF exchange officer, who had previously served on an Aggressor squadron at Nellis AFB.

Late on 2 May the first five Harriers fitted with refuelling probes and long range ferry tanks deployed to St Mawgan. The 100gal combat tanks were carried outboard for later use. The tension was heightened by the news of earlier Vulcan and Sea Harrier attacks on Port Stanley airfield, 8,000 miles away to the south. Sqn Ldr Jerry Pook remembered how daunting it all seemed from St Mawgan.

The first three Harriers flew from St Mawgan early on 3 May, accompanied by Victor tankers. Wg Cdr Squire had expressed concern that the tankers were going to leave them 1,000 miles short of Ascension Island, without navigational assistance or search and rescue cover. A Nimrod was then promised to shadow their arrival, but in the event they were unable to make contact. On the way the tankers were using more fuel than had been planned and announced that part of the formation would have to divert to Banjul in the Gambia. Peter Squire let Tony Harper and Mark Hare go on to Ascension, which they duly reached after a record-breaking sortie of 9 hours 15 minutes. The CO, after refuelling at Banjul with one Victor, joined them at Wideawake Airfield on Ascension late that night, after logging almost 11 hours which included his first night tanking.

The second wave set out on the following morning from St Mawgan, but John Rochfort had to divert into Madeira with a fault. He was collected and brought on by Hercules while Clive Loader flew the sick Harrier back to UK via Gibraltar. The third wave, on 5 May, also ran short of fuel; Ross Boyens dropped into Banjul with a fuel transfer problem and had to stay overnight.

On the next morning six Harriers were flown aboard *Atlantic Conveyor*, together with the eight Sea Harriers of 809 Squadron. Already aboard were seven Naval helicopters and four RAF Chinooks. As Ross Boyens was still flying in from Banjul, Bomber Harris talked Peter Squire into taking him south in place of the young New Zealander. Ross was hopping mad when he heard of the switch, but he was tasked with defending Ascension Island with Tim Smith, in the remaining aircraft. Back at MoD, AVM Hayr had realised the vulnerability of the vital supply link at Ascension; putting the Harriers on alert was his first step in guarding against a possible Argentine commando raid.

Atlantic Conveyor set sail that night for the South Atlantic. The CO, Bob Iveson and the groundcrew were aboard the crowded container ship, while Sqn Ldr Jerry Pook and the other five pilots sailed on the accompanying North Sea ferry *Norland*, which they shared with 2nd Battalion of the Parachute Regiment.

The journey southwards was a time of strange tranquility after the hectic preparations of the previous month. The groundcrew and the FINRAE team worked on, completing the modifications to the aircraft, while the pilots from *Norland* came over each day by helicopter to discuss tactics and to study the latest intelligence reports. Their spirits were raised by hearing of the early combat successes of the Sea Harriers, but a mood of apprehension began to grow as they slipped down the latitudes and the days grew colder. The forthcoming transit flight to HMS *Hermes* would be their first sortie for 12 days and it would cast them into the totally strange environment of a carrier at war. At this point some of them had never actually seen an aircraft carrier!

Atlantic Conveyor's Master, Capt Ian North, became a great friend of the Squadron, and his crew looked after everybody well. 'Cap'n Birdseye' took to conning the ship personally as the weather deteriorated, remaining on the bridge for countless hours to guide her through the heavy swell and minimise the amount of water reaching her precious deck cargo. Chief Tech Fred Welsh, who had been with the Squadron intermittently for over 12 years, celebrated his birthday on board. The crew presented him with an enormous iced cake which turned out to be a disguised case of Carlsberg!

The two ships joined the Task Force on 18 May and prepared the aircraft for the cross-decking sortie to *Hermes*. Two were initially unserviceable, but Peter Squire took off for *Hermes* in the afternoon, followed by Jerry Pook, Peter Harris and John Rochfort.

No 1 Squadron's Harrier GR3s on the deck of *Atlantic Conveyor* together with the Sea Harriers of 809 Squadron. All are encased in tailor-made bags to keep out the salt spray, except for one Sea Harrier which remained on alert during the journey from Ascension Island to the Task Force.
Squadron Archives

The others came over by helicopter with the groundcrew and their equipment. This process took several days, as the carrier and 'Atlantic Convenience' had soon separated and were many miles apart. The last two Harriers were obliged to refuel aboard HMS *Invincible* en route because the vertical take-off from the container ship had greatly restricted their fuel load. One of the last items to be transferred was FINRAE, and this meant that all the early sorties from *Hermes* would have to be flown using the very poor standby instruments with no head-up display.

The pilots remembered vividly that first day on *Hermes*. After a courteous greeting from Capt Middleton, Wg Cdr Squire was treated to a long series of briefings on the way that No 1 would be expected to operate from the ship. He was joined half-way through these by Pook, and later by Iveson, who missed most of the important bits. The CO was then astonished to receive an order to fly three aircraft right away on a dusk training sortie, despite having explained the importance of the missing FINRAE. Squire, Pook and Iveson strapped in on the overcrowded deck and waited apprehensively for the order to launch in the fading light. Their sortie was cancelled, reinstated and finally cancelled again after dark before they were allowed to stand down, firmly convinced that this was the management's little way of putting the 'Crabs' in their place.

After spending the night in their new cramped quarters, they all flew their first carrier sorties on 19 May, armed with AIM-9G Sidewinders, and searched in vain for an Argentine Boeing 707 that had been shadowing the Task Force from outside the Total Exclusion Zone. Jerry Pook later recalled how easy it seemed to fly his Harrier to and from the deck of *Hermes* — the hassles only seemed to start when he left his cockpit and descended into the bowels of the ship.

By now the Sea Harriers of 800 and 801 Squadrons had firmly established their superiority over the Argenine Air Force. They had so far lost only three aircraft out of 20, and none of these had been destroyed in air combat. No 1 Squadron was therefore free to be used exclusively in the ground attack role, despite the absence at this stage of a proper tasking system via a Ground Liaison Officer.

'Mother', or 'the rat-infested rust bucket'. HMS *Hermes* steams through the South Atlantic in the winter sunlight. Sqn Ldr Jerry Pook's photograph shows four of No 1 Squadron's GR3s on deck, together with Sea Harriers of 800 and 809 Squadrons plus a Sea King and a Lynx. *Squadron Archives*

Wg Cdr Peter Squire faces the ski-jump at the start of No 1 Squadron's first offensive operation since 1945. His Harrier is loaded with two BL755 cluster bombs as well as a pair of 100gal drop tanks. Sqn Ldr Jerry Pook prepares to follow him out.
Squadron Archives

Late on 19 May, Wg Cdr Squire received the Squadron's first offensive task since 1945, and the pilots made their final preparations for war.

20 May

The CO decided to lead the first operation himself, together with Sqn Ldrs Jerry Pook and Bob Iveson. FINRAE was still en route from *Atlantic Conveyor*, so they were given a Sea Harrier escort to provide navigational assistance to the target, which was a fuel dump at Fox Bay on West Falkland. 'We'd had the whole morning to plan the sortie in minute detail', said Peter Squire, 'and we produced an immaculate "staff college" solution, which actually worked wonderfully. Of course, it was the only op to get properly planned that way — the rest were all done in a terrible rush in the cramped little ready room.'

Hermes' deck was crowded as never before, with Sea Harriers, bombs, helicopters, Harriers, Sidewinders and all the paraphernalia of war. In the end the Harriers were launched five minutes before their escort and proceeded to the target 30 miles ahead of him, while he called them occasionally to confirm that their radar position was good. The three Harriers swept in from the hills of West Falkland with perfect accuracy. They attacked in a broad vic formation, easing up to drop their Cluster Bomb Units (CBUs) from about 150ft, and using standby sighting information to aim them at the rows of fuel drums. Bob Iveson glanced back as they cleared the target area at very high speed and saw a tremendous series of secondary explosions erupting from the target area. They pulled up to 30,000ft to return to the ship, where they recovered with little difficulty despite their lack of instruments in poor visibility under a low cloudbase.

21 May — D-Day at San Carlos

Long before dawn, the British Marines and Paratroopers began to secure the landing beaches at Port San Carlos. While it was still dark a task was received from the control centre on HMS *Fearless* to attack four helicopters on the ground near Mt Kent. These had been seen by the SAS and they were well placed to insert a counter-attack towards San Carlos. Unknown to the pilots, Mt Kent was a defended area.

'I briefed up Mark Hare, and we strapped in on deck in pitch darkness for our first ever night launch', recalled Jerry Pook.

'We were thankful for the arrival of FINRAE, which meant that we had head-up instruments at last, although it failed to provide us with navigation or weapon-aiming information. We launched in the first glimmer of a clear dawn, and arrived over the islands where it was still pretty dusky. We flew through the grid reference, and saw the helicopters when we were right in the middle of them — a Chinook, two Pumas and a Huey. Thinking the area to be undefended, we raced around the back Mt Kent, and dived in to attack them with CBUs. I pickled mine off a bit low — and missed with the whole bloody lot! Mark's had hung up, so we went round again and strafed them with cannon. Mark got a good hit on the Chinook, and I clobbered one of the Pumas. On the next pass we both hit the second Puma, and we left all three burning well. We then had a go at the Huey, who was air-taxying and evading well, when Mark said, "Hey, I've just been hit". So we ran out to the north, and I could see that he was leaking a bit of fuel. He jettisoned his CBUs in the dusk most spectacularly, and he landed on with about 400lb of fuel and a couple of holes in his wing tank. The boys fixed the aeroplane in about 3 hours and it flew again that afternoon.'

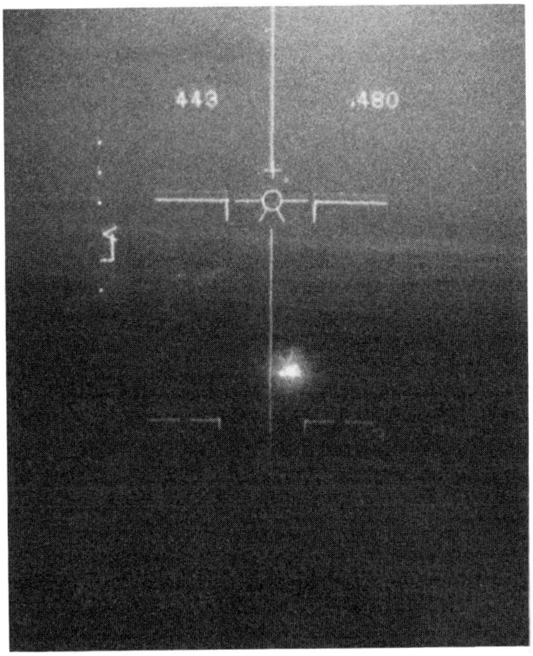

A still from Flt Lt Mark Hare's gunsight Telford camera, taken during the dawn raid on 21 May, shows a helicopter exploding during a strafing pass. The aircraft's pitot boom can be seen at the bottom of the picture. *Squadron Archives*

Later the CO and Jeff Glover launched to work with a Forward Air Controller in the area of the beachhead, but Peter Squire's wheels would not retract and so he sent Glover off alone, feeling that the FAC would help to watch his tail. Unfortunately the FAC had no target for him and he was re-tasked to attack targets at Port Howard, just across Falkland Sound. His first target was too close to the settlement so he did not attack, and returned 15min later from a different direction. Despite his subterfuge he was hit, probably by heavy calibre cannon fire, and ejected into the sea with a broken shoulder. He was rescued by the Argentinians and became the only RAF PoW of the conflict.

Back on the ship the CO felt that he had been wrong to send Glover off alone, and the Squadron did not hear about his survival for several days. But war involves risk, and no amount of peacetime training can compare with the lessons learned through actual combat. Two further armed recce sorties were flown that day, around the beachhead and to Dunnose Head airstrip on West Falkland, but nothing was seen. Harris and Rochfort remembered being very twitched on the way to their target just after Jeff Glover's disappearance.

On 24 May, four aircraft attacked the most heavily defended target of the campaign — Port Stanley airfield. Mark Hare's gunsight picture shows the moment of release of two 1,000lb retard bombs. He is flying at 449kt and 100ft over the runway threshold. A wrecked Pucara can be seen to the left of the runway. The smoke to the right is from Tony Harper's bombs, dropped 30 seconds earlier. *Squadron Archives*

22 May
Shortly after noon, four aircraft were tasked to attack the airfield at Goose Green, whose heavy defences had already claimed Nick Taylor's Sea Harrier on 4 May. Goose Green was the base for a squadron of Pucara light attack aircraft which had already tried to attack the beachhead at San Carlos. To minimise the chance of detection, Pook led them through the hills from the north coast in very poor weather: they ran in low and at the last minute eased up to 150ft to acquire their designated targets. 'I could see no undamaged Pucaras on the strip, so dropped my CBUs on a large camouflaged vehicle', said Jerry Pook. 'I just glanced back when a great burst of AAA exploded across my tail, which was the Oerlikon 35mm that we were afraid of. Bob Iveson saw this as he followed in, and he thought I'd had it!' On this sortie they dropped packets of chaff from their airbrakes to decoy the gun-laying radars. Bomber Harris and John Rochfort attacked a nearby stores area during the same pass; JR scored a hit, but Bomber's CBUs hung up — a repetition of the previous day's problem.

In the evening the CO and Tony Harper went to Weddell airstrip to look for the Pucaras, but found none. En route they spotted the freighter *Monsoonan* in Falkland Sound, but did not attack.

23 May
Just after dawn the Squadron flew the first of several missions to deny the elusive Pucaras the use of various airstrips on West Falkland. In response to Special Boat Service information, Bomber Harris led four aircraft to Dunnose Head where it was decided to crater the grass strip with 1,000lb retard bombs, but no Pucaras or defensive fire were seen. One of Mark Hare's bombs missed the runway and hit a food store in the adjacent settlement; unfortunately an inhabitant received a splinter wound in her bottom, and Mr Tim Miller lost an eye. After the war Mark went to see them and was immediately forgiven — he later became Tim Miller's best man at his wedding in England. The locals had been ready to plough up the strip and were infuriated that the SBS had not asked them to do so.

Pook and Rochfort flew a low-level recce of Port Howard later in the morning, and the film showed a battalion of troops dug in and well camouflaged. In the afternoon the Boss led the same pair to Pebble Island airstrip where they attacked several Pucaras that had already been put out of action in the SAS raid on 11 May.

That night the Squadron received a task from the Captain of *Hermes* to cut the runway at Port Stanley Airfield at dawn with retard bombs, supported by a pair of Sea Harriers for flak suppression. It was pointed out that bombs delivered at shallow angles do not penetrate very well, and Pook suggested that the raid would be much more effective after a reconnaissance sortie to select specific targets. These points were overruled by Capt Middleton, so they sat down with 800 Squadron to plan the mission with the limited information that was available.

Paveway Laser Guided Bombs were used in action for the first time by the Squadron against Argentinian artillery and HQ positions. The fins and guidance kits were assembled in UK after No 1 was deployed to the South Atlantic: they were then parachuted to the Task Force from a Hercules, complete with full DIY instructions.
Squadron Archives

24 May
The six aircraft led by Bob Iveson, launched at dawn, against the most heavily defended target of the war. Knowing that they would be picked up by enemy radar, they descended behind McBride Head while the two Sea Harriers ran in to lob air-burst 1,000lb bombs from outside the defended area. This worked well, and the two pairs streaked in low through the hills to attack from the west while the heavy AAA was looking east. But there was still plenty of tracer as Iveson and Harper bombed the edge of the runway, and the CO and Mark Hare also had to cope with the debris of the earlier bombs. The retard bombs had been incorrectly fused and had bounced off the runway momentarily before exploding. Squire and Hare were both hit several times by small-arms fire but returned safely to the ship, angry about the circumstances that had

caused them to risk their lives to no great effect. Later reconnaissance showed that the runway had been hit by at least four bombs. These had caused minimal damage, but a helicopter and possibly three Pucaras were disabled by bomb fragments.

25 May — Argentina's National Day

The Squadron went back to Port Stanley airfield three times on 25 May, but stayed clear of the defences by toss-bombing from low level. The Harriers bombed from vic formation, led by a Sea Harrier which measured the range with great accuracy on its Blue Fox radar. Roland and Tigercat SAMs were fired against them, but all fell short.

The pilots were sitting in the ready room on *Hermes* in the afternoon when they heard of the loss of HMS *Coventry* in an air attack. Some time later *Hermes* herself declared an imminent air attack. Peter Squire remembered the apprehension as they sat and listened to the running commentary from the bridge as a Super Etendard raid was picked up on radar: 'They detected the Exocet launch, and we knew that it was intended for us. The chaff rockets started going off, and there was a great bang as one of *Invincible's* Sea Darts homed on to our chaff and exploded over the flight deck.' Jerry Pook had his fingers in his ears, but still heard another dull explosion:

'The Commander said that they'd hit *Atlantic Conveyor*, and she was on fire. We were still at "State Zulu", but I managed to sneak up through the Met Office where I could see her burning well, enveloped with smoke, and with all the helicopters picking up survivors. Sadly, dear old Captain North was not among them — he'd been such a good friend to us on the way down.'

26 May

Early in the morning Bob Iveson and Tony Harper attacked the troop concentration at Port Howard that Pook had discovered three days before, going in with cluster bombs and inflicting heavy damage. Later, a search for artillery around Mt Kent revealed nothing, and one aircraft was sent to lob another three bombs at the main airfield.

Just before dark Pook and Hare searched the line from Teal Inlet to Wireless Ridge. Close by Mt Kent they found another Puma which Pook destroyed with CBUs. Then near Two Sisters they were 'locked up' by a Super Fledermaus radar and a Blowpipe was fired at Pook, fortunately without causing any damage.

27 May — The Battle for Goose Green

After a further loft bombing sortie to the Port Stanley runway, urgent tasks began to be received from *Fearless* in support of 2 Para's pitched battle north of Goose Green. The weather was very poor and Bob Iveson and Mark Hare were scrambled after several cold hours of cockpit alert on deck. The FAC was unable to bring them in — he had chosen a run-in point that was hidden by cloud, and he had no Laser Target Marker (LTM) to help them — but Iveson found a gap in the clouds and dived down to attack. He picked up the target very late, but was able to talk Hare in for a successful pass. Knowing that the Paras were in trouble, Bob made two further attacks — greatly against his better judgement — and struck a company position under heavy flak before escaping to *Hermes* to refuel. The pair were re-armed quickly and sent straight back to Goose Green, where the Oerlikons were practised and waiting.

'Back over the islands nothing had changed. This time we were asked to attack another camouflaged gun, but on the first run neither of us saw the target at all. I therefore elected to go for another company position which we both attacked with CBUs, and then we climbed out en route for *Hermes*. At this point I made my final tactical mistake, by telling the ALO that we still had cannon available. I can still hear myself saying those words, and I will never really know why! I think it was the euphoria at having escaped the flak again,and as a quip to show that I hadn't been as frightened as I really was. Unfortunately the ALO had just had reports of troop movements along a track near Darwin, and please could we go and strafe them. I could feel poor Mark's icy glare on the back of my neck as we ran in towards the well-practised AAA guns. Having nearly emptied my guns on the Argentine troops in a shallow dive, I ran out across Goose Green airfield, and finally received my come-uppance!'

Bob Iveson's Harrier was hit by two 35mm shells in quick succession. The nose began to drop and, with flames in the cockpit, he ejected horizontally from below 100ft.

'It was quite exciting — but everything must have worked as advertised. I hit the ground with a hell of a thump and knocked myself scatty. After releasing my 'chute, my main problem was that I couldn't see properly.

Sqn Ldr Bob Iveson was shot down over Goose Green by enemy AAA on 27 May. Despite minor injuries, he evaded capture and hid in Paragon house overnight before being rescued by an Army helicopter. *Squadron Archives*

Looking up, I saw what appeared to be men approaching, so I pushed off as fast as I could in the opposite direction. When my vision began to come back I realised that they must have been sheep!'

Having dumped his survival pack in his haste, Bob hid from a searching helicopter and then found shelter in a nearby empty farmhouse. Being of a hungry disposition he was delighted to find a full larder and a comfortable bed, where he settled down for the night.

Meanwhile, Jerry Pook was tasked by the CO to reconnoitre Goose Green's defences in the dusk, but really to listen for Bob's SARBE beacon. Orbiting at 15,000ft he thought that he was out of range of the guns, but when a burst of heavy calibre tracer flashed past he retreated to a safe distance. He then watched the bomb splashes at Ajax Bay as enemy Skyhawks attacked the refrigeration plant where the British field hospital had been set up. He was too far away to intercept, and nothing was heard of Iveson.

28 May

The CO led three aircraft to attack a storage area on Mt Kent with rockets and cannon in the morning, under a cloudbase of about 100ft in visibility of 2-3km! They survived this and were overjoyed to hear Bob Iveson's faint call sign on the radio. Peter Squire's request for a rescue helicopter was rejected in case Bob had been captured and the call was a spoof.

Above Goose Green the battle was reaching its climax following the loss of Lt-Col 'H' Jones. The exhausted Paras were within sight of the settlement, but they were being held up by direct fire from the 35mm Oerlikons based on the small pensinsula by the airstrip — the same guns that had claimed Taylor's and Iveson's Harriers. Maj Chris Keeble, who had taken command of 2 Para when his Colonel was killed, called for an immediate airstrike as the weather closed in. Bomber Harris, Tony Harper and Jerry Pook were scrambled from *Hermes* and arrived as the light was beginning to fade. 'When we got there, 2 Para had just about run out of steam', said Pook.

'The Argies were pinned down on the peninsula, and the battlefield was covered in the fog of war. This time the FAC worked well. We went straight in and walloped the guns with CBU and RP. It was a strangely pretty attack in the dusk with everything going off. This time they did not shoot back.'

The attack came at a crucial moment in the battle. The fighting stopped as darkness fell, Maj Keeble was soon able to negotiate the surrender of Goose Green, and the shell-shocked defenders gave in to a force of Paras half their number. Keeble signalled the Squadron to say that Harris' raid had made a most significant contribution to the battle.

30 May

After a day of terrible weather on 29 May, the Squadron launched five separate pairs in support of the advance across East Falkland. These included a sortie called by the SAS against possible radars at Mt Round and another against a company position on Mt Kent.

Later, Harris and Harper attacked Port Stanley airport using Paveway Laser Guided Bombs (LGBs) for the first time. Harper dropped his bombs in the direction of the runway while Harris tried to steer them in from 30,000ft using his Laser rangefinder. Unfortunately, this did not work.

The introduction of Paveway is another example of the extraordinary improvisation that characterised Operation 'Corporate'. Back at MoD, AVM Hayr had organised a trial that proved that the Harrier could use the LGB. Kits comprising the nose and tail fins and the fragile guidance system were packaged, and were tested for 'dropability' by pushing them off a 10ft high wall. They were then parachuted to the Task Force from a Hercules. The kits included full instructions on how to strap the fins to a normal thousand pounder, so as to produce a DIY laser guided bomb!

At 14.30hrs Pook and Rochfort were sent to attack a helicopter site near Mt Wall. While running in Pook felt a thump, and 'JR' called to tell him that he was leaking fuel. The helicopters had gone so they switched their attack to a nearby artillery battery, which Rochfort destroyed with a salvo of rockets. After firing his own rockets Pook ordered them home:

'By now I was streaming fuel, the radio packed up and I had a hydraulic failure. At the top of climb, the gauges started going down really fast, and I realised I wouldn't make the ship. Unbeknown to me JR had called "Mayday" on my behalf — I wasn't sure how far we were from the ship, and I was a bit panic-stricken. Eventually the gauges reached zero, the engine wound down, I waved goodbye to JR, glided to 10,000ft — and pulled the handle. The chute opened, and as I descended through cloud, I could hear rotor blades! A Sea King from 846 Sqn picked me up, and took me home'.

Unknown to Pook another helicopter was approaching *Hermes* at the same time, carrying Bob Iveson. They met below deck: 'Hello Bob, I hear you've been on a walking holiday', said Jerry, 'And you, I gather, have just completed your annual dinghy drill' replied Iveson.

Both pilots proclaimed their fitness to fly, despite sore necks and backs suffered during ejection. But there were now only three aircraft to fly, so they were given the rest of the day off!

31 May
After a further abortive attempt to steer laser guided bombs at the Port Stanley runway, the Squadron received an immediate task from the bridge to attack some 'swept-wing aircraft' that had been spotted at the airport by a passing Sea Harrier from *Invincible*. These might just have been Super Etendards, so the Captain ordered Wg Cdr Squire and Mark Hare to scramble, with no chance to plan or brief with the two Sea Harrier pilots who would be providing flak suppression. 'It was a bit hot and frantic on the run-in', said the CO, who quickly picked up the targets. 'I had my blinkers on as usual, fastened on to the target, but Mark could see blokes out on the balcony of the control tower firing at us with rifles.' Both aircraft were hit by small-arms fire and debris, but they launched their 2in RP at the suspicious targets. These turned out to be previously-damaged Aeromacchi jet trainers, which had been visible on No 1 Squadron's recce photos for the past week.

This is a good example of how No 1 Squadron was at times misused during Operation 'Corporate'. It was now firmly allocated in support of the developing land battle, and yet was often tasked on these random sorties. Some tasks received from Wg Cdr Fred Trowern on board the command ship *Fearless* were over-ruled in favour of those generated on board *Hermes*. Despite the friction that this caused at command level, the Squadron's relationship with the carrier was generally excellent, and Flt Lt Brian Mason's 18 groundcrew received magnificent support from the ship's air engineers.

1 June
The sole serviceable GR3 searched for enemy targets between Bluff Cove and Goose Green. Flt Lt Ted Ball, on exchange with 800 Squadron provided top cover, supposedly from 10,000ft because the Sea Harriers were too valuable to be risked to ground fire. Being a good crab he bent his authorisation in order to cover the CO's tail properly from low level. Squire attacked a possible mobile radar site on Mt Usbourne and his film later showed some Argentine activity at Fitzroy Ridge.

Later in the day Flt Lts Murdo Macleod and Mike Beech arrived on *Hermes* with two replacement Harriers from Ascension, after a 3,500-mile sortie lasting 8hr 15min. This feat of outstanding airmanship was accomplished with the aid of eight Victor tankers, and the only possible diversion on route was the heli-deck of RFA *Engadine*, which was in transit half-way. After the last 'prod' their tankers had turned for home, leaving two lonely little jets to navigate the South Atlantic for the final two hours, unsure of the weather, to seek out a speck called *Hermes* for their first deck landing.

2-4 June
A spell of appalling weather now clamped down on the Task Force and the carriers withdrew to the east, out of range of further air attack. This provided a much-needed opportunity for rest, repair and re-briefing. On

shore at Port San Carlos the Royal Engineers were completing a runway of alloy MEXE-planks, 850ft long, plus a landing pad and taxiways. This was smaller than planned, because much of the planking had gone down with *Atlantic Conveyor*. Sqn Ldr Sid Morris, another Harrier pilot from the early 1970s, commanded the forward operating base. 'Sid's Strip' would enable Harriers and Sea Harriers to refuel in mid-operation and also to wait on alert much closer to their targets.

5 June
Operations resumed when Bob Iveson, still sore from his ejection, led Tony Harper to attack targets near Mt Kent:

'We were launched without head-up instruments into thick fog, with no chance of landing back on the ship in 50 yards visibility. Having recovered from the launch, just, I found that I had no IFF — not good for penetrating the Rapier and Sea Wolf screen at San Carlos! In the event we failed to contact our FAC, so we headed for the strip. We also failed to contact the local anti-aircraft controller to tell him about my lack of IFF. In the end we crept round Fanning Head with gear down and lights on, to be locked up immediately by friendly radars — but fortunately not shot at, and we landed safely at the strip'.

Harris and Rochfort later took these aircraft back to the ship, while Iveson and Harper remained ashore for two days.

Pook and Hare launched later to try to find the ground-launched Exocet trailer which the Admiral's staff reckoned would be hidden somewhere on the south coast between Port Stanley airfield and Bluff Cove. 'We felt that this was a pretty good time to get blown away', said Jerry Pook, 'so I got Mark to spoof the AAA and SAMs around Stanley by popping up just out of their range while I came in from the other direction.' Pook streaked through without a scratch, but no Exocets were seen on the film. They were later discovered in the town itself, having scored a hit on *Glamorgan* on 12 June.

Later that day Wg Cdr Squire led a pair that attacked troops in the Two Sisters area, and another pair did some more 'gardening' at Pepple Island airstrip with 1,000lb retard bombs in case the Pucaras returned.

6 June
On another day of poor weather the CO led Mike Beech to look again in vain for the ground-launched Exocets. He was able to drop his CBUs on some positions revealed by Pook's film and Beech cut his teeth against a small battlefield surveillance radar.

7 June
'The day started with a right shambles on deck', remembered Pook.

'Two pairs were tasked separately for FAC and close air support, but Bomber and Tony Harper went u/s, so JR was eventually joined up with me after much faffing about by the Fishheads. My FAC had no trade for us, but luckily I was carrying the position of the Sapper Hill artillery with me.'

They came in low through the hills for minimum exposure, but had to attack into the pale sun. Pook was hit again and holed, and a passing Sea Harrier saw SAMs fired at both aircraft, but the Harriers returned to the carrier without trouble.

Bob Iveson and Tony Harper landed back on *Hermes* from the strip at last light. While ashore they had watched from a slit trench in awful fascination as a Sea Dart from *Exeter* shot down a reconnaissance Lear Jet in clear skies at 40,000ft. Iveson's back was now hurting badly, and he could not bear to pull more than 3G. He was evacuated home by BP tanker on 11 June, where a fractured vertebra was diagnosed.

8 June
The second of two pairs was led to Sid's Strip by the CO. He suffered an engine wind-down when coming to the hover and did well to smash his aircraft down on the strip, tearing some of it up in the process. Mark Hare landed on what was left, and flew back to the ship later. The Boss's aircraft was badly damaged, but he escaped with a shaking.

Two further Harriers were flown to *Hermes* from Ascension by Nick Gilchrist and the long-suffering Ross Boyens. This was an even longer flight, of 3,900 miles, and with no diversion whatsoever available. These aircraft had been fitted with the ALE-40 chaff and flare dispensers, as well as an ingenious I-band jammer in one gun pod to fox the AAA gun-laying radars.

9 June
Two pairs were launched to attack artillery positions at Mt Longden and Sapper Hill which were threatening to hold up the remorseless advance by the Marines and the Paras on Port

Stanley. These positions always seemed to be well-defended, and both missions were engaged with AAA and small arms fire. Murdo Macleod, flying his first operational sortie, was hit in the belly and had to blow his wheels down with the emergency system before landing on *Hermes*.

10 June

The first pair flew straight to the strip at dawn to await tasks from the land forces. After a long wait a target was received, but the weather was poor and the pilots returned to the ship. The CO and Mark Hare then made individual photo runs in front of the advancing troops in the hills to the west of Stanley. Both were heavily shot at but they returned with good pictures of various camouflaged positions. The Boss later returned to attack some of these with Macleod, who took another small-arms hit, this time in the windscreen.

By now the Squadron was ready to use laser guided bombs in conjunction with laser target markers wielded by chosen Forward Air Controllers. A trial was arranged for 18.00hrs, but *Hermes* reorganised its flying programme, despite protests from Wg Cdr Squire. The Harriers were launched two hours early, but nobody told the FAC — so the pilots were obliged to jettison their valuable bombs and return to the ship, seething with frustration.

11 June

Tony Harper led a pair armed with laser guided bombs to contact an SAS patrol at first light. The patrol had been holed up opposite Port Stanley for some days and the leader was keen to hit a police house near the edge of town that was being used as a headquarters. Sadly, the batteries in his laser target marker had run down, so the bombs could not be guided. Exhibiting great faith in No 1's weapon aiming skills he then invited Tony to toss the bombs at the target unguided! Harper took one look at the civilian houses close by and lobbed his bombs in the direction of Wireless Ridge instead.

That afternoon John Rochfort had a lucky escape on take-off when the wind speed over the deck was much lower than that stated by the ship. He sank horribly after leaving the ski-jump, and had to jettison all weapons to miss the sea by a few feet. He accompanied Ross Boyens to provide mutual support, and they strafed troop positions on Mt Longden. When running in to the target they used the new ALE-40 chaff and flare dispenser for the first time.

In the evening the CO and Hare made a daring raid on artillery and headquarters positions on the eastern slope of Mt Tumbledown, near Moody Brook barracks, well within the fire zone of the airfield defences. They came in fast, at below 50ft, and popped up to bomb. Peter Squire was hit in the cockpit on the run-in and the bullet missed him by inches. Because of an arming error, all their bombs fell 'slick', without the retard parachutes. Thankfully they also failed to explode under the Harriers, and skipped on towards Stanley. SAMs were fired at both Harriers as they sped away, one following the CO's aircraft but falling short. Another Blowpipe round caught up with Mark Hare. He saw it above him, took violent evasive action very close to the ground, and the SAM exploded harmlessly just above his canopy. Moody Brook barracks had to be evacuated that night because an unexploded bomb was stuck in the wall!

12 June

The day's operations were concentrated on Sapper Hill. Harris and Macleod were the second of three pairs to attack artillery positions behind the hill and, like the others, they had to fly through heavy AAA and small arms fire to reach their targets. Murdo took his customary hit, which punched a hole in his rear reaction control duct. As he decelerated to the hover back at the ship, hot air from the duct started a fire in his tail section. He landed quickly, and just in time, for his control runs were starting to melt.

The battle damage teams below were now working flat out. All 18 groundcrew were highly experienced; between them they achieved marvels of improvisation, often effecting repairs of drop tanks and wiring looms that would have been rejected as impossible by a Maintenance Unit. Once again the inherent strength of the lightweight Harrier continued to amaze them all.

Later in the day Tony Harper took Nick Gilchrist on his first operation to Sapper Hill. They caught a large number of troops in the open, walking along a road, and devastated them with CBUs. Harper's aircraft was hit by two rifle bullets and AAA splinters during their single pass attack.

13 June

As British forces closed the net around Port Stanley the FACs were able to take up good vantage points with their laser target markers. In mid-morning Wg Cdr Squire arrived behind the front line of troops, armed with two

Paveway LGBs. His FAC illuminated a company HQ and the CO ran in from the west and lobbed his first bomb high above the target. However, the FAC turned his kit on too soon and the bomb exploded into the peat some 400yd short. The second bomb guided perfectly and blew the target to pieces. Squire returned jubilantly to the ship, never having exposed himself to enemy fire.

Four hours later Jerry Pook arrived, similarly armed, and his first bomb scored an absolute direct hit on a 105mm gun in a camouflaged sangar on Mt Tumbledown. He too had remained behind the front line, but this did him no good, for he came under heavy fire from the Scots Guards. They had seen the release of his first bomb and had assumed that they were themselves under attack! Fortunately their aim was no better than their recognition.

14 July
Before dawn, British forces began their final assault on Port Stanley. The Squadron's first pair was deployed to the strip to wait for targets from the Army.

At mid-day Bomber Harris and Nick Gilchrist strapped into their two Harriers on the deck of *Hermes*, each carrying two Paveway LGBs. They climbed to a position behind the front line and waited for their final brief. It never came. The FAC was lining up his laser on a likely position when the inhabitants raised a white flag and began running towards Port Stanley. Wg Cdr Fred Trowern, at Brig Thompson's headquarters on Mt Kent, told the Naval Liaison Officer to call off the attack, and the Harriers were sent back to the ship, bearing the first news of the Argentine surrender.

Back on *Hermes* No 1 Squadron's aircraft were rapidly re-armed with Sidewinders and joined the Sea Harriers on combat air patrol, lest the surrender on the Islands resulted in vengeance from the mainland. This never happened — and the war just petered out.

The Harriers were flown ashore on 4 July, to commence air defence operations from the devastation and squalor of Port Stanley airfield. Here two Harriers stood on alert with the pilots strapped in to sub-zero cockpits for four hours at a stretch, although they were later allowed to hold five minutes readiness from some Argentine cabins in six to seven hour shifts.

The original No 1 Squadron pilots and groundcrew were replaced as quickly as possible, initially with crews from RAF Germany. They came home by Hercules and by VC-10, slumped exhausted among the freight and the troops, to a hero's welcome at Brize Norton. Last to come was Flt Lt Jeff Glover, released to Montevideo by the Argentinians. No 1 Squadron had survived intact.

The last man to come home. Flt Lt Jeff Glover steps off a VIP Devon at Wittering after his release. After being shot down on his first mission, he was taken to Argentina with a broken shoulder as the only RAF PoW. After being well treated, he was released to Montevideo, and brought home by VC10.
Squadron Archives

For their leadership and courage Wg Cdr Peter Squire and Sqn Ldr Jerry Pook were awarded the Distinguished Flying Cross, while Bob Iveson and Mark Hare were Mentioned in Dispatches. Peter Harris and Mike Beech later received Air Force Crosses, while Sid Morris got the OBE and Brian Mason the MBE.

Later in the year No 1 sent a further detachment to Port Stanley to maintain the air defence of the Falklands for several months. As the Squadron's diarist put it:

'By the end of the year, No 1 (Fighter) Squadron has ploughed the Atlantic good and proper; East to West and back, North to South and back, and back, and back. Still, we were given seven days off at the end of it all'.

Epilogue

Marshal of the Royal Air Force Sir Dermot Boyle GCB, KCVO, KBE, AFC, RAF, presented No 1 Squadron's second Standard on 21 June 1983, with the following words:

'I have it in command from Her Majesty the Queen to present you with your new Standard. It is a great privilege for me to do so. I still remember with nostalgia the time I spent in No 1 Squadron in Iraq 57 years ago.

'Your Squadron is unique in the Royal Air Force not only because it is No 1, but in addition it can trace its origins back to 1878 through No 1 Company of the Air Battalion of the Royal Engineers . . . You are also privileged at the present time to be operating an aircraft which with vectored thrust incorporates a unique British invention which has made the Harrier a war-winning weapon.

'Your Squadron has operated in many theatres throughout the world, but surely your journeys to Ascension Island and then the Falklands battle zone must be one of the most spectacular events in your long and distinguished history . . . You flew some 130 operational sorties in the battle zone in ground attack, close support and reconnaissance missions against heavily defended targets and caused severe disruption to enemy command centres, fuel and weapon storage dumps, artillery positions and airfields . . . Your squadron performed with courage and distinction.

'On these achievements I warmly congratulate you — as I do on the smartness and efficiency of this parade.

'You have demonstrated to a remarkable degree the far-reaching versatility of the air weapon, and I present you with your new Standard confident that you will honour and protect it and enhance the traditions and achievements which it represents; confident that you will strive to prove worthy of your ambitious motto — *In All Things First*'.

Right:
A pair of Squadron Harriers over northern Norway.
Squadron Archives

Commanding Officers

Maj E. M. Maitland	13 May 1912
Maj C. A. H. Longcroft	1 May 1914
Maj W. G. H. Salmond	28 January 1915
Maj P. B. Joubert de la Ferte	19 August 1915
Maj G. F. Pretyman DSO	24 November 1915
Maj G. C. St P. de Dombasle	24 December 1916
Maj A. Barton-Adams	20 June 1917
Maj W. E. Young	3 August 1918
Sqn Ldr J. B. Graham MC, AFC	18 March 1920
Sqn Ldr G. G. A. Williams	10 November 1922
Sqn Ldr E. O. Grenfell MC, DFC, AFC	8 October 1923
Sqn Ldr E. D. Atkinson DFC, AFC	25 May 1924
Sqn Ldr C. N. Lowe	19 April 1926
Sqn Ldr E. D. Atkinson DFC, AFC	11 April 1927
Sqn Ldr E. O. Grenfell MC, DFC, AFC	19 March 1928
Sqn Ldr C. B. S. Spackman	27 July 1931
Sqn Ldr R. W. Chappell MC	21 November 1933
Sqn Ldr C. W. Hill	1 October 1934
Flt Lt T. R. McEvoy	31 January 1936
Sqn Ldr C. W. Hill	1 December 1936
Sqn Ldr F. R. D. Swain AFC	12 April 1937
Sqn Ldr I. A. Bertram	15 January 1938
Sqn Ldr P. J. H. Halahan DFC	17 April 1939
Sqn Ldr D. A. Pemberton DFC	24 May 1940
Sqn Ldr M. H. Brown DFC	10 November 1940
Sqn Ldr R. E. P. Brooker DFC	23 April 1941
Sqn Ldr J. A. F. Maclachlan DSO, DFC	3 November 1941
Sqn Ldr R. C. Wilkinson OBE, DFM	31 July 1942
Sqn Ldr A. Zweigbergk	30 May 1943
Sqn Ldr J. Checketts DSO, DFC	3 April 1944
Sqn Ldr H. P. Lardner-Burke DFC	29 April 1944
Sqn Ldr D. G. S. R. Cox DFC	11 January 1945
Sqn Ldr R. S. Nash	21 April 1945
Sqn Ldr H. R. Allen DFC	9 January 1946
Sqn Ldr C. H. Macfie DFC	26 October 1946
Sqn Ldr T. R. Burne DSO, DFC	15 July 1947
Maj R. Olds USAF	4 February 1949
Sqn Ldr T. R. Burne DSO, DFC	1 October 1949
Maj D. F. Smith USAF	10 January 1950
Sqn Ldr J. L. W. Ellacombe DFC	18 August 1950
Sqn Ldr R. B. Morison DFC	21 November 1952
Sqn Ldr D. I. Smith	27 July 1953
Sqn Ldr F. W. Lister DSO, DFC	1 December 1953
Sqn Ldr R. S. Kingsford	8 August 1956
Sqn Ldr L. de Garis AFC	5 July 1958
Sqn Ldr J. J. Phipps	1 December 1958
Sqn Ldr P. V. Pledger	1 January 1961
Sqn Ldr F. L. Travers-Smith	1 January 1963
Sqn Ldr D. C. G. Brook	28 December 1964
Sqn Ldr G. Jones	1 November 1966
Sqn Ldr L. A. B. Baker	20 September 1968
Wg Cdr J. A. Mansell	10 April 1969
Sqn Ldr L. A. B. Baker	21 May 1969
Wg Cdr D. Allison	4 August 1969
Wg Cdr K. W. Hayr AFC	1 January 1970
Wg Cdr E. J. E. Smith OBE	6 January 1972
Wg Cdr P. P. W. Taylor AFC	3 December 1973
Wg Cdr J. G. Saye	9 July 1976
Wg Cdr R. B. Duckett AFC	17 July 1978
Wg Cdr P. T. Squire DFC, AFC	26 March 1981
Wg Cdr J. D. L. Feesey AFC	23 December 1983
Wg Cdr I. M. Stewart	13 June 1986

When operating in winter on the northern flank of NATO, the Harriers' normal grey and green camouflage scheme is lightened with whitewash to make the aircraft blend with the snowy landscape.

Squadron Bases

Location	*Date*
Farnborough	May 1912-August 1914
Brooklands	August 1914-November 1914
Netheravon	November 1914-February 1915
St Omer	February 1915-March 1915
Bailleul	March 1915-March 1918
St Marie Cappelle	March 1918-April 1918
Clairmarais	April 1918-August 1918
Fienvilliers	August 1918-October 1918
Senlis	October 1918-October 1918
Bouvincourt	October 1918-November 1918
Le Hameau	November 1918-March 1919
London Colney	March 1919-October 1919
Uxbridge	October 1919-January 1920

Disbanded and simultaneously re-formed in India

Risalpur, India	January 1920-May 1920
Bangalore, India	May 1920-April 1921
Hinaidi, Iraq	April 1921-November 1926

Returning to England

Tangmere	February 1927-September 1939
Octeville	September 1939-September 1939
Norrent-Fontes	September 1939-October 1939
Vassincourt	October 1939-May 1940

Retreating through France via Berry-au-Bac, Condé, Anglure, Châteaudun, Rouen, Angers and Nantes, May 1940-June 1940

Location	*Date*
Tangmere	June 1940-June 1940
Northolt	June 1940-September 1940
Wittering	September 1940-December 1940
Northolt	December 1940-January 1941
Kenley	January 1941-April 1941
Croydon	April 1941-May 1941
Redhill	May 1941-June 1941
Kenley	June 1941-June 1941
Redhill	June 1941-July 1941
Tangmere	July 1941-July 1942
Acklington	July 1942-February 1943
Biggin Hill	February 1943-March 1943
Lympne	March 1943-February 1944
Martlesham Heath	February 1944-April 1944
North Weald	April 1944-April 1944
Predannack	April 1944-June 1944
Harrowbeer	June 1944-June 1944
Detling	June 1944-July 1944
Lympne	July 1944-August 1944
Detling	August 1944-December 1944
Manston	December 1944-April 1945
Coltishall	April 1945-May 1945
Ludham	May 1945-July 1945
Hutton Cranswick	July 1945-April 1946
Tangmere	April 1946-August 1956
Nicosia	August 1956-December 1956
Tangmere	December 1956-June 1958

Re-formed from No 263 Squadron at Stradishall

Stradishall	July 1958-December 1961
Waterbeach	December 1961-August 1963
West Raynham	August 1963-July 1969
Wittering	July 1969-

Postscript

An account of its activities during the Falklands War is a fitting end to the illustrious history of No 1 Squadron. Of course, the Squadron has remained active since its return from the South Atlantic, with a round of deployments and exercises. The accompanying photographs are just a small sample to illustrate some of the tasks and events undertaken by the Squadron since 1982.

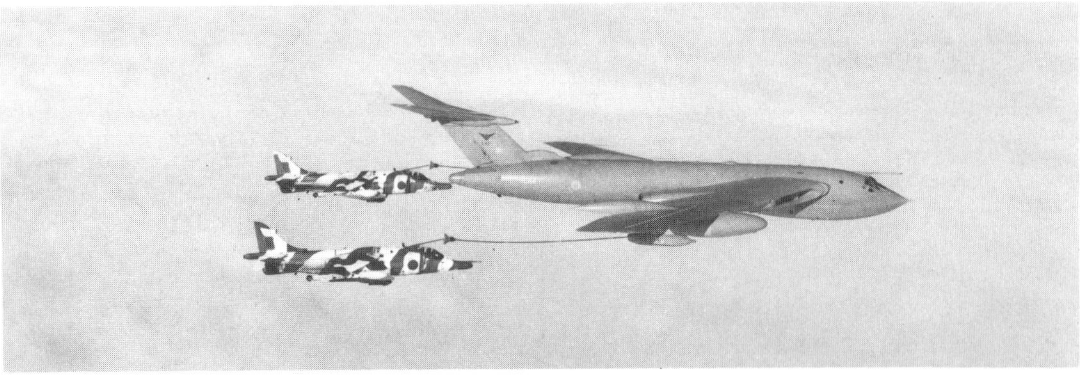

Below:
Deploying to Tromso in northern Norway for Exercise 'Anchor Express' on 5 March 1986. Two of No 1's Harrier GR3s refuel from a Victor tanker of No 57 Squadron. *MoD*

Bottom left:
In May 1985 No 1 Squadron regained its sea-legs with a deployment on board HMS *Invincible* in the Western Approaches. *Squadron Archives*

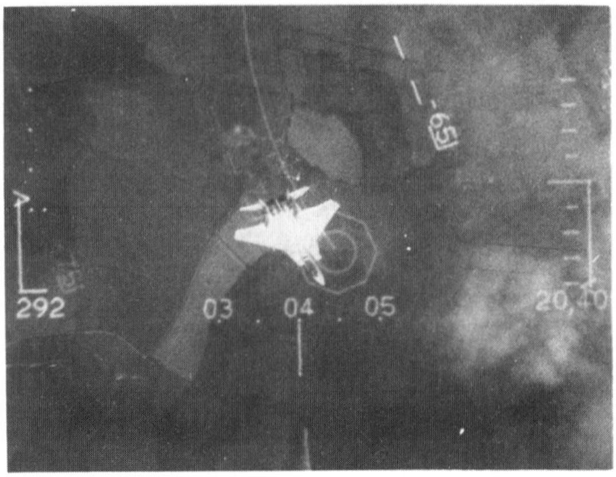

Bottom right:
Air combat training often includes dogfighting with formidable foes. Here a No 1 Squadron pilot achieves a killing camera shot on a USAF F-15 at 20,000ft. *Squadron Archives*

Index